Research in Biotechnology

Principles of Experimental Design in Biotechnology

Rock Canyon High School

Volume 1

April 2016

Editors

Shawndra L Fordham

Bryan M Winkelman

Cover Photo Credits:
Staphylococcus aureus bacteria escape by NIAID/RML [Public domain], via Wikimedia Commons
Sunflower oil and sunflower by www.torange.us [CC BY 4.0], via Wikimedia Commons
Egg is protein by By Pdcook [Public domain], via Wikimedia Commons
2489 Blue jeans [Public domain], via pdpics.com
S-N-S Flame leg by Glenn Woodell (http://www.windvisions.com/me.html) [CC BY-SA 3.0], via Wikimedia Commons
CRISPR-Cas9 File:NHGRI-97218 by Ernesto del Aguila III, NHGRI [Public domain], via Wikimedia Commons
C. elegans by wormbase.org [used by permission], via WormClassroom.org

Additional photos, as well as all interior photographs and visuals, taken or created by the authors.

Cover designed by Shawndra L Fordham

ISBN: 1530478596
ISBN-13: 978-1530478590

ACKNOWLEDGMENTS

The success of this inaugural year of Rock Canyon High School's Principals of Experimental Design in Biotechnology has been the result of a combined effort of so many people, from our generous research funders, to our amazing mentors, and everyone in between. I want to extend a special thank you to Bryan Winkelman for teaming with me in making this class a success. This class wouldn't be the same without all of your contributions, including your innovative ideas, design of the website, and help with publication of our journal and scientific posters. I also want to thank Amy Hacker and David Ferguson, fellow science teachers at RCHS, for helping the students design and conduct their research. Both of you have supported the students with technical aspects of their research and have given your valuable planning time to help them; I cannot thank you both enough. In addition, I am grateful to Jason Dunkle and Matthew Gracey, fellow math and science teachers respectively at RCHS, for providing support to the students as they performed statistical analysis on their data. I am excited to be partnering the class next year with Wendy Lerolland's technical writing class, making the class a more interdisciplinary experience for the students. I am grateful to the RCHS Department of Science and the school administration for supporting this course, as well as the Douglas County School District for providing the Innovative Education Grant, along with granting the program Perkins Grant money to fund the purchase of research-grade science equipment the students used this year in their research.

I would finally like to thank all of the families, friends, and other donors who contributed to the students' projects. The students will thank you personally in their individual acknowledgments, but I want to recognize the following donors who donated $100 or more to the students' research this year:

Tom Bogard
Rob Burkholder
Patrice Isabella
Diane M. Keely
Hong Kim
Adele and Joseph Merkle
Robert Merkle
Kristen Schurr
Mark Tobey

CONTENTS
April 2016

Foreword

FOREWORD

My vision for this research course, Principles of Experimental Design in Biotechnology, was for students to conduct authentic scientific research from start to finish. This included designing their research experiments, finding an expert mentor in the field, writing and defending a research proposal, writing a budget and securing their own sources of funding, ordering their materials, conducting their research, writing a full text journal article and publishing it in a journal, making a scientific poster, and presenting their research at a research symposium. Students were as excited, and as nervous, as I was when we started, but somehow they trusted me enough to take a risk and try something new. They are so incredibly brave to join me in this adventure without any of us knowing what we were in for or capable of accomplishing.

The two biggest hurdles we initially had to overcome were finding mentors and funding. I was surprised at the reaction of professionals in the field when we started reaching out to get mentors. Our amazing scientific community stepped up and willingly gave their time to these young researchers. The willingness of professionals -- scientists and researchers -- to help these kids design their investigations, supply materials, and be a source of support and professional knowledge is unbelievable. As you read through the students' articles, you will find all mentors recognized for their time and support. Although the biggest hurdle turned out not to be a major challenge this year, every year will be different, and I am excited about the relationships we have developed.

The students' efforts to secure funding was another turning point for the class. Without a budget to support their research, nothing could be accomplished. With the help of Bryan Winkelman, our Teacher Librarian, the students created their own website to communicate their research proposals and secure funding (bit.ly/rchsrb2). Students blogged about their research to keep people updated on their progress throughout the year, and they were funded quickly.

As the students began ordering their materials and conducting their experiments, I was astonished at the change in atmosphere in what I used to think of as "my" classroom. The classroom had been transformed into the students' space, and I transitioned from their teacher to their mentor and source of support. Students came and went in lab throughout the day as their research dictated. Although I struggled to let go of control and let students take care of business, I began to develop a trust in them that I never had as an instructor prior to this experience. Several times I was moved to tears watching them come in and get to work without a word spoken from me. When they needed me, they let me know. When they had questions, they asked. Otherwise, I was expected to stay out of their way and let them work. Having these experiences was transformative for all of us. They learned to take control of their learning and I learned to let go of that control.

As we end the year preparing to publish and designing the scientific posters, the students are struggling. They are used to doing their assignments and getting a grade back. Now, they must do revision after revision only to have their work given back to them for yet one more revision to create an article worthy of publication. At the end of their senior year they are frustrated. Yet it is at this point where I am seeing the students learn the most valuable lesson. Their perseverance is being tested and they have created this amazing publication. We have emerged from this experience despite facing many challenges, surprises, and frustrations, having accomplished everything we set out to accomplish and more. I am so proud of these students and I am excited for the future of this class.

Shawndra L. Fordham
Biotechnology Teacher
Rock Canyon High School
April 24, 2016

Species of bacteria associated with leg amputee residual limbs

Allison M Kerker, Emily E Sattem, and Shawndra L Fordham

Department of Science, Principles of Experimental Design in Biotechnology, Rock Canyon High School, Highlands Ranch, Colorado, USA

Amputees wear liners and sleeves over their residual limbs that hold prosthetic sockets on and prevent them from moving or sliding. These liners and sleeves prove to be quite a problem when it comes to skin irritations and patient satisfaction; amputees often get harsh skin reactions on their residual limbs possibly due to increased bacterial colonization under the liner/sleeve. This research aims to identify the species of bacteria found on amputees' residual limbs and to compare those findings to the species of bacteria on non-amputees. This was accomplished by swabbing both amputees and non-amputees and extracting the DNA to be sequenced and compared. Ideally, this data will contribute to the introduction or alteration of antibacterial practices in the prosthetic field. After research was conducted, it has been found that _Staphylococcus_ and two species of _Corynebacterium_ are significantly more prevalent on amputees, and _Streptococcus_ is significantly more prevalent on non-amputees. _Staphylococcus_ and _Corynebacterium_ are both bacteria that are associated with various skin infections; therefore, it can be speculated that these bacteria may be the cause of amputee specific skin irritations.

Currently, there are upwards of two million people coping with amputation in the United States, and many of them develop painful skin irritations on their residual limb that prevent them from performing their normal daily activities. An average of 185,000 people lose a limb per year in the United States alone. The top causes of limb loss are vascular disease, diabetes, trauma, and cancer, respectively.[2] This study sought to determine if there are different species of bacteria on amputee residual limbs in comparison to non-amputee knees, which could be tied to the cause of these skin irritations. The tightness of the sleeve over an amputee's residual limb creates prime conditions for bacterial growth. It has been found that many amputees develop rashes or other skin irritations, which lead to numerous challenges and increases the difficulty of daily tasks. This research sought to identify the species of bacteria present with the hope that there could be specific antibacterial treatments implemented to decrease the number of patients who develop rashes and irritations in the future.

A prosthetic device includes any artificial body part that replaces a missing or nonfunctional part. The most common prostheses are artificial limbs that can be created specifically for a certain patient's arm or leg regardless of whether the limb is missing below or above the elbow or knee. Modern day technology is constantly increasing the applications for prostheses. The advancements of 3D printers, 3D carvers, and digitizers have made amputees' lives drastically simpler. Digitizers are used to computerize an amputee's unique residual limb shape and size, by simply scanning a cast of the stump. Then, 3D printers and 3D carvers can be manipulated to create basic sockets out of foam with minimal human involvement.[4]

Plastic polymer laminates are most commonly used to create prosthetic limb sockets. The polymer begins as a liquid and is combined with a hardening agent which binds the fabrics to each other, creating a strong and lightweight product that is also flexible. Once cooled, the plastic polymer laminate maintains its form well. Also, certain laminates, such as polyester and epoxy, can be made thinner and lightweight in certain areas, while thicker and stronger in other areas where more pressure is put on the socket. A disadvantage with this material is the fact that it is not as easy to alter with heat after original molding.[21] Due to this construction (trapped perspiration, limited airflow, etc.) of prosthetic sockets, the most common skin bacteria become more prominent.[6]

Staphylococcus epidermidis recently has become the most common cause of hospital contracted infections. It is most commonly found on the skin, but is the leading cause for

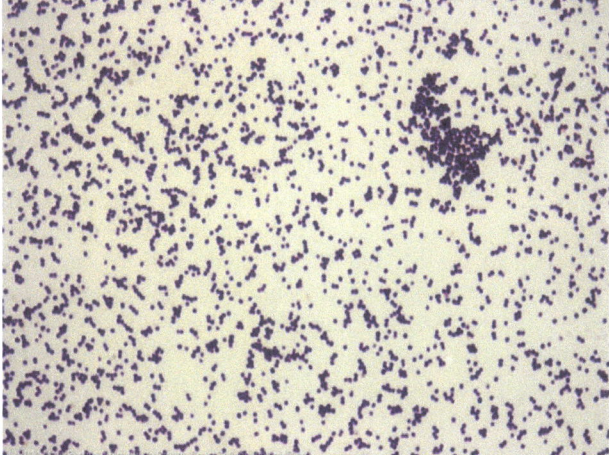

Picture 1: _Staphylococcus epidermidis_ is the leading cause of disease resulting from medical implants.[15]

disease resulting from implanted medical devices/prosthetics (**Pic. 1**). This particular species of *Staphyloccocus* has a specialized genome that allows it to survive in environments with extremely high salt concentrations and high osmotic pressures, such as the environmental conditions found on amputees' residual limbs.[18] A similar bacteria, *Staphylococcus aureus* is the leading cause of all *Staphylococcus* infections. These infections are contracted from contact with surfaces or people who possess the bacteria.[12] An additional microorganism group associated with skin infections is *Micrococci*, which can vary drastically genome-wise. *Micrococci* are arranged in clumps called tetrads. *M. luteus* is the most common *Micrococci* found on the surface of human skin and breaks down human sweat into odorous compounds (**Pic. 2**). The ideal conditions for *Micrococci* growth include small amounts of water, high amounts of salt, and human body temperature.[16] Furthermore, *Streptococci* is a bacteria that is grouped in pairs and chains and normally found on the skin. Most strains are harmless, but group A and group B *Streptococci* result in diseases such as scarlet fever.[14]

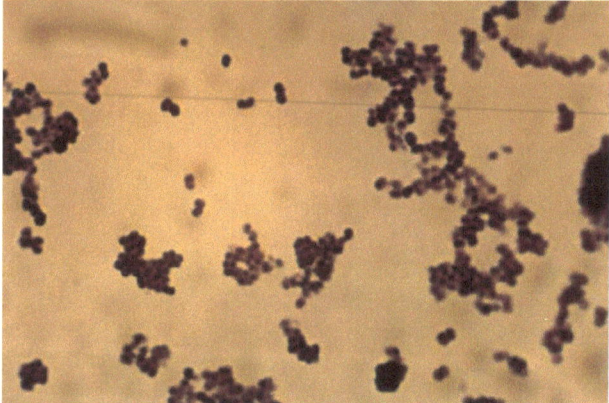

Picture 2: *Micrococcus luteus* breaks down sweat, creating odor, and is the most common species of *Micrococci* found on human skin.[12]

Over the last 25 years, there have been a handful of studies done on the microbial growth associated with the residual limbs of amputees. However, they each vary in terms of what was being investigated. Dr. Kathryn Buikema and Dr. Jon Meyerle (2014) conducted research with the Department of Dermatology in Maryland that investigated the most common skin complications on amputee residual limbs and the possible causes for these complications. This research, and others, have determined the wetness of the residual limb to be the leading factor in bacterial growth and the cause of residual limbs' thin, extremely fragile, and easily broken skin. As a result, this skin has difficulty healing under the constant pressure and wetness of the environment created by the prosthetic socket. The most common skin complications found in this study were ulcers, dermatitis, acroangiodermatitis, folliculitis, dyshidrotic eczema, tinea manuum, and tinea pedis.[5] Another experiment, by Kohler, Lindh, and Bjorklind (1989), investigated the types of residual limb bacteria as well as the effects and amputee satisfaction with a cleaning antiseptic. One thing they noted was the natural bacteria found on healthy skin were also the main types of bacteria

found in their samples. These included *Staphylococcus aureus*, *Staphylococcus epidermidis*, diphtheroids, and alpha-hemolytic *Streptococci*.[10]

After we swabbed leg amputee residual limbs and the skin above non-amputee knees, we extracted and amplified the DNA with bacterial primers provided by Dr. Noah Fierer's lab at the University of Colorado Boulder. With their help, we used high-throughput sequencing and the Fierer Lab's database, known as BaseSpace, in order to determine the species of bacteria present in each species. BaseSpace connects to the Illumina sequencer to help store and analyze the sequencing data. It has cloud storage that connects other Illumina products and shares data. Our sequences were compared against the expansive database associated with BaseSpace.[3] Using this, the type of microbial growth for each sample was then identified and analyzed.

METHODS

In this study, the species of bacteria inhabiting leg amputee residual limbs were compared to those inhabiting the skin just above non-amputee knees. This was accomplished by swabbing leg residual limbs of amputees and above the knees of non-amputees, then extracting the bacterial DNA from those swabs and sequencing it. This study was kept anonymous by assigning each participant a sample letter or number. Amputee swabs were labeled and will be referred to as a number 1 through 20. Non-amputee swabs were labeled and will be referred to as a letter A through T. The results from the two groups (amputees and non-amputees) were compared on 13 different axes in order to analyze the data.

Participants

20 amputees and 20 non-amputees were swabbed. They varied in age and included both males and females. Participants were swabbed at either the Adaptec Prosthetic Clinic in Littleton, CO or Rock Canyon High School in Highlands Ranch, CO.

Procedure

In order to determine the species of bacteria present on the residual limbs, we first obtained the swab samples from leg amputees at Adaptec Prosthetics in Littleton, Colorado. We used dry, dual-end swabs from Fisher Scientific. The dual-end swab allowed us to have a backup sample in case we encountered a problem and needed to re-run the sequence. After collection, we had two samples from each amputee. In order to standardize the approximate amount of biomass collected on the swab, we swabbed a three-inch by three-inch area of skin located on the bottom of leg amputees' residual limb. In order to have a control group, we also swabbed the same area one inch above the knees of people who do not have a leg amputation. This allowed us to determine the species of bacteria typically found on an average human leg. Furthermore, we asked the patients to not shower or put on lotion during the morning they were swabbed to ensure proper collection of microbes and to prevent the collection of antibacterial soap, which could interfere with the DNA extraction. We obtained samples during mid-afternoon, as close to 2 P.M. as possible.

Next, the DNA was extracted from each sample using the PowerSoil DNA Extraction Kit from MoBio **(Pic. 3)**. Detailed protocols outlined in this kit were followed and

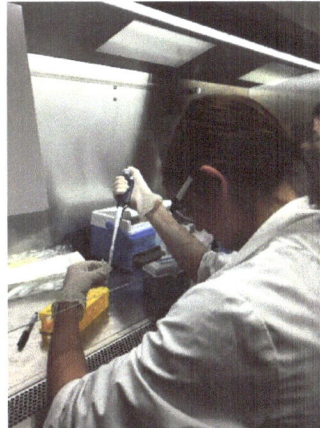

specific alterations were provided by Robin Hacker-Cary, our mentor who is a student at the University of Colorado Boulder in Dr. Noah Fierer's lab.[17] These alterations included initially incubating the frozen samples at 60°C for 10 minutes, vortexing for 2 minutes as opposed to 10 minutes, putting samples on ice as opposed to incubating them at 4°C, and storing the samples in the refrigerator overnight at the halfway point, which was immediately after

Picture 3: As part of the extraction protocol, Emily is removing the supernatant and leaving the pellet at the bottom.

adding solution C4. Our samples were then stored at -20°C until we had collected and extracted all of the samples. We did three extraction runs and all extractions were performed in the Labconco BioSafety Cabinet (model 302681100) at Rock Canyon High School **(Pic. 4)**.

Polymerase Chain Reaction (PCR) was conducted next. Each well in the plate had 10.5 µL of water, 12.5 µL of Promega GoTag Hotstart Colorless Master Mix, 1µL of gDNA, and 1 µL of primer. The bacterial primer was a combination of the Illumina 5' Adapter (AATGATACGGCG ACCACCGAGATCT

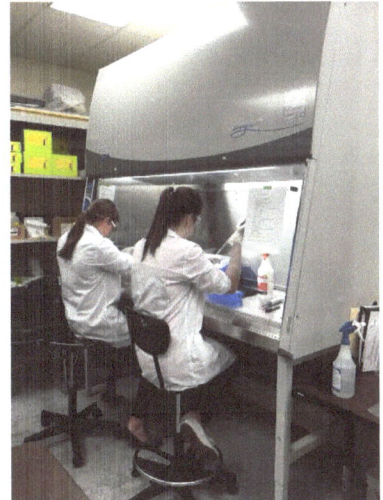

ACACGCT), the golay barcode (12 nucleotides that is

Picture 4: In order to maintain a sterile environment, all of the extractions were performed in the Labconco BioSafety Cabinet.

different for each sample), the forward primer pad (TATGGTAATT), the forward primer linker (GT), and the 515f forward primer (GTGYCAGCMGCCGCGGTAA) where M is any nucleotide, which varies depending on the species of bacteria. The thermocycler was set to run for 34 cycles of 94°C for 3 minutes, 94°C for 45 seconds, 50°C for 60 seconds, 72°C for 90 seconds, and then held at 4°C once the process was complete.

We then ran gel electrophoresis to verify that PCR worked. A 2.0% DNA agarose gel was loaded and ran at 135 volts. Afterwards, we examined the DNA fragments on the gel using a UV light **(Pic. 5)**.[1]

We then performed SequalPrep to prepare the samples for sequencing.[8] SequalPrep standardizes the amount of DNA

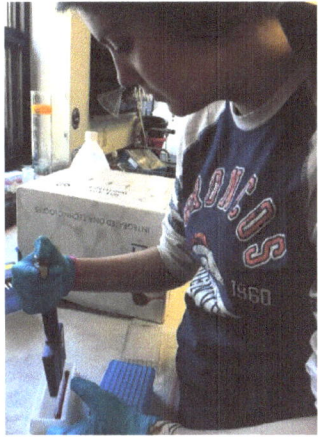

in each sample and then pools all of the samples into a single tube which was stored at -20°C, for 38 days, until ready for sequencing using MiSeq with a 97% confidence rating.

PCR, gel electrophoresis, SequalPrep, Miseq, and data analysis all took place at the Fierer Lab at the University of Colorado Boulder.

Our sequence data was compared to the data in the Fierer Lab's database. We then made an Operational Taxonomic

Picture 5: Allison is pipetting loading buffer with a pipette that is capable of pipetting eight samples at once. This buffer coats the DNA with a negative charge so it will run on the gel.

Unit (OTU) table, performed t-tests for the major OTU's, and analyzed the data using Excel to compare the microbial communities, determine what species of bacteria are statistically more prevalent, and determine the richness of each with percent relative abundances.

RESULTS

When we ran gel electrophoresis on the 47 swab extractions, all of them amplified with proper banding at the expected base pairs. This indicates that DNA was successfully extracted for all 47 swabs. Some samples had brighter bands than others, indicating more DNA extracted, but there was no significant trend regarding the banding.

Clustering of Amputee and Non-Amputee Groups

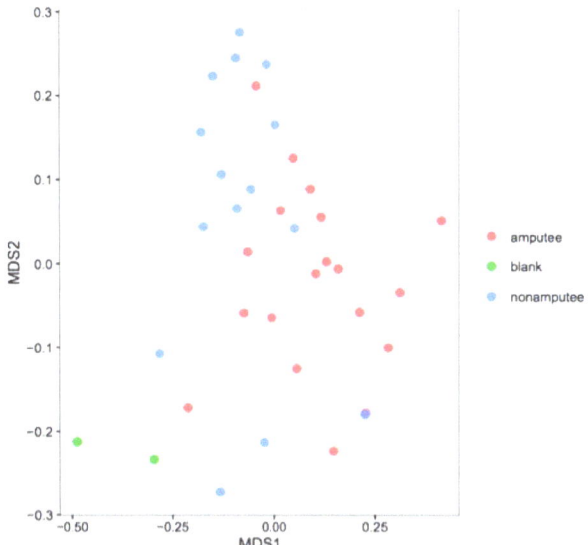

Graph 1: This dissimilarity matrix shows the clustering of the amputee swabs compared to that of the non-amputee swabs.

Graph 1 is a dissimilarity matrix of our data. The dissimilarity matrix shows the relative comparison of the two samples (amputee and non-amputee) based on two factors, taxonomy and relative abundance. The bacterial species are weighted based on relative abundances where those with greater relative abundances are assigned a lower distance than those species that are less abundant. The x and y axes are measured in arbitrary values that combine relative abundances and taxonomy. Each dot represents a single sample. For example, one of the blue dots is Non-Amputee B. Basic clustering can be seen on the matrix; the non-amputee (blue) dots are towards the left, and the amputee (red) dots are towards the right. The two blank (green) dots represent the two PCR blanks. They have relatively low values and are very similar to one another but are not too similar to the amputee and non-amputee samples. This indicates that there was little contamination in the PCR process and that the results for the non-amputee and amputee swabs are reliable.

When we got the OTU table back with all of our relative abundances, *E. coli* was present in all 47 samples. This indicates some sort of contamination that was most likely caused by the equipment used during DNA extractions. In order to derive the desired results, we subtracted the *E. coli* results from each sample.

T-Tests for the Nine Most Abundant OTUs			
Species	**OTU**	**P Value**	**Confidence Interval**
Staphylococcus	OTU_2	0.0017	7.62337 to 30.15379
Corneybacterium 1	OTU_4	0.0395	0.563493 to 21.495005
Corneybacterium 2	OTU_6	0.0127	0.87228 to 6.80888
Corneybacterium 3	OTU_222	0.3763	-3.58794 to 9.25803
Acinetobacter johnsonii	OTU_12	0.7297	-10.43638 to 7.37980
Streptococcus	OTU_8	0.0324	-7.89077 to -0.36655
Corneybacterium 4	OTU_3788	0.7843	-3.41911 to 4.49408
Corneybacterium 5	OTU_15	0.1376	-0.81299 to 5.65261
Finegoldia	OTU_14	0.1891	-4.36994 to 0.89588

Table 1: This data table shows the results of the t-tests that were run on the nine most abundant species of bacteria in the DNA extractions. A t-test compares the relative abundances in the amputee swabs to that in the non-amputee swabs. The p value indicates statistical significance if it is under 0.05, and a lower score indicates a lower probability that the results were obtained through chance.

Table 1 shows the results of the t-tests we ran on nine OTUs. These nine OTUs were chosen because they had the largest relative abundances and these abundances varied greatly between amputees and non-amputees. The bacterial species associated with these nine OTUs were: *Staphylococcus*, five different species of *Corynebacterium*, *Acinetobacter johnsonii*, *Streptococcus*, and *Finegoldia* **(Graph 2)**.

The MiSeq sequencer was programmed to identify DNA sequences by comparisons to other sequences in the BaseSpace database at a 97% confidence rate. As a result, the majority of our OTU's came back only identified to the genus level. Even though each OTU is a different species, the sequencer was not able to identify the species. In order to differentiate each species within the same genus, we have

assigned matching genus names a number to indicate the separate species.

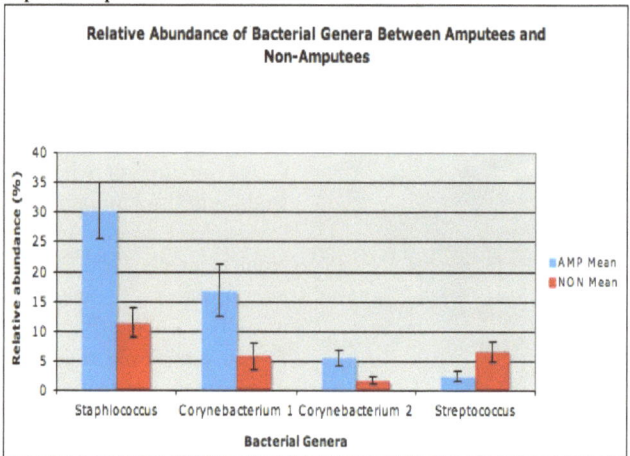

Graph 2: This graph compares mean relative abundance for amputees (blue) and non-amputees (red) obtained from the t-tests for each specific species. The black bars indicate the standard mean error, which shows how similar the data was in each sample group for that species.

DISCUSSION

This study aimed to compare the species of bacteria present on leg amputees' residual limbs to those present above the knees of non-amputees. Similar research to this was done in 2008 by Dr. Erol Koc and Dr. Ercan Arca. They published a descriptive study on the skin problems in amputees. They examined 142 amputees for dermatological problems while also taking note of age, gender, age at time of amputation, incident that led to the amputation, level of amputation, and type of sleeve (soft socket versus silicone). They found that 27.1% of the patients using soft sockets and 36.4% of the amputees using silicone didn't have any skin complications; soft sockets had greater statistical significance for dermatological problems (82.9%). Of the patients that were tested for bacterial growth, approximately half tested positive for skin irritations caused by bacteria. The most common bacteria found were Group A B-hemolytic *Streptococcus*, *Staphylococcus aureus*, and *Staphylococcus epidermidis*, respectively.[9] From this, it was hypothesized that these species of bacteria would be found on the amputees' residual limbs. In addition, we expected to find a species that was relatively more abundant on residual limbs than on the skin above non-amputees' knees.

In the end, the statistically significant genera in terms of relative abundances on amputee residual limbs were found to be *Staphlococcus*, *Corynebacterium* (species 1), *Corynebacterium* (species 2), and *Streptococcus*. The low p-values (all below 0.05) indicate a high statistical significance between the amputee and non-amputee sample groups since it indicates a low probability that our data occurred by chance alone. Therefore, the most statistically significant difference in amputee and non-amputee relative abundances was OTU_2, which is a species of *Staphylococcus*. *Staphylococcus* is more present in amputees than it is in non-amputees. Also, the standard error means for each species do not overlap when comparing amputees to non-amputees. This indicates that the relative abundances for each sample group were compact and similar within a group but not too similar to the other sample group, which also indicates statistical significance.

The most common *Staphylococcus* complication is skin infection. These skin infections are characterized by hot, swollen, painful red areas of the skin. *Staphylococcus* is harbored in open wounds. Unfortunately, amputees often get cuts on their residual limbs from contact with the bottom of their sockets, and this could be a reason as to why they have more *Staphylococcus* than non-amputees, who would not normally have a cut above their knee.[18] Typical and reliable treatments of *Staphylococcus* are antibiotics, wound drainage, and device removal. Removing the prostheses and draining the wound would cease the growth and spread of the *Staphylococcus* and allow the infected skin to recover and heal.[11] The majority of *Staphylococcus* infections can be prevented with regular and thorough cleansing. In this case, the leg amputee's liner and residual limb should be cleaned at least daily. Rubbing alcohol would be best on the liner as long as it is completely dry before coming in contact with the skin. For the residual limb, gentle antibacterial soap would be best. This would thoroughly cleanse the residual limb while not stripping the skin and making it more vulnerable to bacteria.[13]

Corynebacterium is most commonly found in wet, warm locations, such as under a prosthetic liner, where perspiration is trapped. Although harmless in most cases, *Corynebacterium* can cause severe skin irritations that can spread to and infect the lungs and airways. The most common *Corynebacterium* complication is diphtheria, which is a contagious infection of dead epithelial cells. *Corynebacterium* is easily spread in hospitals. Amputees spend a lot of time in clinics, VAs, or traditional hospitals, so it makes sense why they would have more *Corynebacterium* than non-amputees.[15] In most cases, *Corynebacterium* can be killed with antibiotics. It is most sensitive to penicillin and penicillin equivalents, such as rifampin and fluoroquinolones.[20]

Streptococcus can induce mild skin irritations that can normally be treated with topical cream. When ingested, *Streptococcus* can cause other severe conditions such as pneumonia and blood infections.[14] Many species of *Streptococcus* are not pathogenic, and since the data doesn't indicate which species of *Streptococcus* was present in large quantities on the residual limbs, we cannot conclude if the *Streptococcus* is causing the amputees' skin irritations. Nonetheless, treatment for *Streptococcus* skin infection is antibiotics as well. Penicillin is reliable, but if allergic, erythromycin and azithromycin are just as reliable.[7]

Our experiment is not free of errors: On the first DNA extraction (the first set of 16 swabs), the incubation step was skipped. We immediately notified our mentor and took our completed extractions up to the Fierer Lab at the University of Colorado Boulder to be tested for the presence of DNA. Fortunately all 16 extractions showed thick banding on a presumptive gel, which indicates that they, in fact, had DNA, despite us missing a step. On the second extraction run (the next 16 swabs), there were four

samples that were transferred and then their tubes were discarded, so the number/letter of them were unknown. Three of them were able to be salvaged, but unfortunately, sample M had to be thrown out and was not used in the study. An additional error occurred during DNA extractions. All of our samples, except for the two blank PCR runs, came back with evidence of *E. coli*. This was clearly a contaminant as it is not commonly found on human skin and was present in our controls. It can be concluded that it occurred during DNA extractions at Rock Canyon High School because the two blanks were not added until PCR and the subsequent steps were performed, which occurred at the Fierer Lab; the PCR blanks contained no evidence of *E. Coli* contamination. *E. Coli* is often used in the Rock Canyon High School Laboratory, and we did not autoclave or use barrier tips, so the contamination is most likely from a non-sterile micropipette. However, the same micropipettes were used for all of the extractions, so each sample was equally affected by contamination. Thankfully, the two PCR blank runs were added after extractions, which allowed us to determine that *E. coli* could be removed and ignored during data analysis. It was simply a contaminant that did influence the other OTU data and was not factored in.

To extend this research, future researchers should obtain International Review Board (IRB) approval to conduct a survey of all participants when collecting swab samples. This would allow researchers to investigate if there is a link between the species of bacteria found and the participants' gender, age, activity level, substance use, lifestyle, and type of socket worn. This could lead to discovering a behavior that is linked to a harmful species of bacteria that is the cause of amputees' skin irritations.

Another way future researchers could extend this research is by swabbing a residual limb and above the knee on the same participant. The majority of leg amputees are only missing one of their legs, as opposed to both of their legs. An amputee swab and a non-amputee swab could be taken from the same participant. For example, if an amputee has his left leg amputated, but his right leg remains, his left residual limb and above his right knee could both be swabbed. This would make comparing the two groups much more reliable. When comparing an amputee swab and non-amputee swab from the same person there is no longer the possibility that one person does something to contract different species of bacteria than another person.

ACKNOWLEDGMENTS
We would like to give a huge thank you to Royce and Rae Heck and all of the employees at Adaptec Prosthetics. This project would not have been possible without you and your continued assistance and support. Thank you for helping us contact amputees and getting all 20 swabs by our deadline. We would like to thank Robin Hacker-Cary and the Fierer Lab at CU Boulder for allowing us to do lab work in their facility and for providing us with primers, PCR, gel, SequalPrep materials, and sequencing free of charge. We cannot thank Robin enough for being in constant contact with us and helping us with the entire process. We would like to thank all of the participants who allowed us to swab their knee or residual limb. We never knew how easy it would be to

find 40 willing people to feature in our research. Thank you for your sample! We would like to thank all of those who donated money towards our funding and supplies including Tom Bogard, Rob Burkholder, Nicholas Laatsch, Kristen Schurr, Lori Dishneau, Mark Grafitti, and Margaret Koperny. A very special thanks to Kathleen Steffe, Donna Sattem, and Chris Bagnell for donating! We would like to thank Matthew Gracey for helping with our data analysis and Tom Dillon for giving us feedback on our research proposal. We would like to thank Andy Abner, Rock Canyon High School, and Douglas County School District for providing us with supplies, workspace, funding, assistance, and unwavering support to make this research possible. We would like to thank Susanne Petri and Amy Hacker for allowing us to use their room space and equipment. We would like to thank Sheri Bryant for fully funding two Biosafety Cabinets for the biotechnology program at Rock Canyon through the Perkins grant. We appreciate the opportunity to perform the majority of our lab work in the biosafety cabinet, as it was crucial to maintain a sterile environment during extractions. We would like to thank Bryan Winkelman for showing enthusiasm in our research and for reading through our posts and papers. Also, thank you for always being available and flexible.

REFERENCES

1. Addgene. (2013). Agarose Gel Electrophoresis. PubMed. Retrieved 2015, October 8. [Web]
2. Amputee Coalition. (2015). Limb Loss Statistics. National Health Council. Retrieved 2015, October 8. [Web]
3. BaseSpace. (2016, March 21). Sequence Hub. Illumina. Retrieved 2016, March 28. [Web]
4. Bradford, T., & McGimpsey, G. (2006). Limb Prosthetics Services and Devices. Retrieved 2015, October 8. [Web]
5. Buikema, K. E.S., Meyerle, J.H. (2014). Amputation stump: Privileged harbor for infections, tumors, and immune disorders. *Clinics in Dermatology*, 32, 670-677.
6. Davis, C. (1996). Normal Flora Medical Microbiology. Samuel Baron. Retrieved 2015, October 8. [Web]
7. Davis, C. P. (2016, February 19). Group A *Streptococcus* Infections. Stoppler M. C. Retrieved 2016, April 13. [Web]
8. Invitrogen. (2008). SequalPrep Normalization Plate (96) Kit. Invitrogen. Retrieved 2015, October 8. [Web]
9. Koc, E., Tunca, M., Akar, A., Hakan Erbil, A., Demiralp, B., and Arca, E. (2008). Skin problems in amputees: a descriptive study. *International Journal of Dermatology*, 47, 463-466.
10. Kohler, Lindh, and Bjorklind. (1989). Bacteria on stumps of amputees and the effect of antiseptics. *Prosthetics and Orthotics International*. 13, 149-151.
11. Mayo Clinic Staff. (2014, July 11). Diseases and Infections: Staph Infections. Mayo Clinic. Retrieved 2016, April 13. [Web]
12. Medline Plus. (2014, July 25). Staphylococcal Infections. Medline Plus. Retrieved 2015, October 8. [Web]
13. Medline Plus. (2016, April 5). Staph Infections. Medline Plus. Retrieved 2016, April 13. [Web]
14. Medline Plus. (2014, September 3). Streptococcal Infections. MedlinePlus. Retrieved 2015, October 8. [Web]
15. Microbe Wiki. (2015, July 22). *Corynebacterium*. Microbe Wiki. Retrieved 2016, April 8. [Web]
16. Microbe Wiki. (2010, August 6). *Micrococcus*. Microbe Wiki. Retrieved 2015, October 8. [Web]
17. MoBio. (2013). Soil Kit Protocol. MoBio. Retrieved 2015, October 8. [Web]
18. Otto, M. (2010, August 1). *Staphylococcus epidermidis* – the "accidental" pathogen. NCBI. Retrieved 2015, October 8. [Web]
19. Perkins, M. (2011, November 27). *Staphylococcus epidermidis*. NCBI. Retrieved 2016, February 28. [Web]
20. Soriano F., Zapardiel J., Nieto E. (1994, October 31). Microbial Susceptibilities of *Corynebacterium* Species and Other Non-Spore-Forming Gram-Positive bacilli to 18 Antimicrobial Agent. American Society for Microbiology. 39, 208-214.
21. Uellendahl, J. (1998, November 1). Prosthetic Primer:Materials Used in Prosthetics Part II. NCBI. Retrieved 2015, October 8. [Web]

ABOUT THE AUTHORS

Pictured (Top): Emily (left), Allison (middle), and Royce Heck (right). Mr. Heck is one of our mentors. He and his partner Clint own Adaptec Prosthetics, a prosthetic clinic in Littleton, Colorado. Adaptec helped us to align 20 amputees to swab. **Pictured (Bottom):** Emily (left), Robin Hacker-Cary (middle), Allison (right). Robin was our mentor at CU Boulder. As an undergraduate student, she helped us set up our project, work through the protocols, and interpret our data.

This process has been anything but simple. However, among all of the challenges, we have learned many valuable skills. The most beneficial thing that we learned is how to be resilient and moreover, how to stay calm and fix mistakes rather than getting frustrated. Scientific research is never perfect and there are always mistakes to be made. There were a few times where, what seemed like a minor mistake, jeopardized our entire research. In these moments, we had to quickly and effectively think of a solution, without letting the problem get any worse. From this we learned the value of being able to handle and recover from errors.

On another note, the lab experience we gained was invaluable. As we head off to do undergraduate research next year (Emily will be attending Kansas University and Allison will be attending the University of Minnesota), simply being in a lab, following sterile protocols, and effectively communicating is extremely beneficial for us. We are more confident when it comes to executing lab work, following lab protocols, and working in a college level lab.

Lastly, the people that we have met this year, whether it be participants or mentors, have greatly impacted us. The connections we have made with mentors and other scientists in the field will be helpful in future scientific endeavors.

Assaying the effects of Namzaric and coffee on paralysis in beta-amyloid peptide Alzheimer's disease model *Caenorhabditis elegans,* strain CL2006

Allie N Kellner, Brooke M Galyon, and Shawndra L Fordham

Department of Science, Principles of Experimental Design in Biotechnology, Rock Canyon High School, Highlands Ranch, Colorado, USA

In this research project, the effects of coffee and a pharmaceutical drug, Namzaric, were tested to identify if these substances had a statistical impact on the rate of paralysis in Alzheimer's disease model, *Caenorhabditis elegans (C. elegans)*, strain CL2006. These model organisms were transformed with the human Alzheimer's disease gene, pCL12, causing them to produce beta-amyloid peptides (Aβ), and resulting in neuronal plaques associated with Alzheimer's disease that cause paralysis in *C. elegans*. By exposing these worms to coffee and Namzaric over a lifelong period of 12 days, the results of the rate of paralysis were compared between control and treatment plates. In order to observe our plates, we picked 6 adult worms to each plate at the same time and allowed them to lay eggs for 2 hours, then picked off the adults to create a synchronized population. From there we counted the number of non-paralyzed, paralyzed, and dead worms at the same time daily and recorded our results over the 12 day period. A two proportion z-test was used to compare the coffee and the Namzaric treatment data to the control. It was concluded that both coffee and Namzaric slow the rate of paralysis in the sample populations in comparison to the control group receiving no treatment according to the z-scores, p̂-scores, and p-values gathered from the data. All z-scores were above 3 and all p-values were <0.0001, meaning the variance between our control and experimental treatments are statistically significant. In addition, it was found that there was no statistical difference between the two experimental treatments. These findings show that while both Namzaric and coffee have a significant effect on the toxicity of Alzheimer's disease by decreasing the rate of paralysis, there is no statistical difference between the effects coffee and Namzaric have on the rate of paralysis in *C. elegans* strain CL2006.

A chronic neurodegenerative disease, Alzheimer's Disease (AD), is suspected to be caused primarily by the accumulation of insoluble β-amyloid peptide (Aβ) between neurons in addition to neurofibrillary tangles of tau protein. Such aggregation of Aβ is what results in the formation of plaques associated with AD in humans. Aβ protein aggregation between neurons also inhibits communication between synaptic gaps in neurons, resulting in cell death. This neurodegeneration causes the human brain to lose a significant mass of brain tissue, severely inhibiting human functioning. AD is a late onset disease which is associated with progressive symptoms such as impairments in cognition and memory. Throughout this experiment, we used two different chemicals, Namzaric and coffee extract, to test the effect they have on Aβ expression caused by AD and the subsequent effects on paralysis of *C. elegans* when they are present in the growth media throughout the lifespan of the organisms. According to the Alzheimer's Association, AD is the 6th leading cause of death in the U.S., affecting 5.3 million people in America alone.[6] The Alzheimer's Association also claims that of the top ten causes of death, it is the only one that cannot be prevented, cured or slowed. The symptoms and progression of AD vary between individuals. This is also true when

researching the disease in model organisms.

For our research, we used *Caenorhabditis elegans (C. elegans)*, a model organism used in studying genetics in relativity to humans. *C. elegans* were first introduced as a model system by Sydney Brenner in 1963. According to Corsi, Wightman, and Chalfie, *C. elegans* are microscopic nematode ring worms that are found worldwide.[2] These nematodes reproduce quickly, are inexpensive, very small (1mm in length), have short life cycles of approximately 23 days, self-fertilize (as they are hermaphrodites), and have a completely mapped cell lineage. Because of this, they are great organisms to be used to study human genetics and age related diseases such as Alzheimer's.

For our experiment, we used *C. elegans* strain CL2006 obtained from Dr. Christopher Link at CU Boulder. These nematodes were co-injected with a transgene that includes the human Alzheimer's gene pCL12 (*Punc-54*:: SP::Aβ 1–42) as well as the rol-6 gene which results in a rolling phenotype.[7] Within the model organisms, this transgene causes the expression of Aβ and results in paralysis of the *C. elegans*. In strain CL2006, Aβ accumulates in the constitutive muscle tissue of the *C. elegans*, ultimately resulting in the paralysis of the worms because the motor neurons reach a point where they are no longer able to communicate as Aβ accumulation interferes in the synapse of neurotransmitters, inhibiting action potential in the neurons to take place and cause locomotion. By creating synchronized populations of *C. elegans* and exposing them to coffee, Namzaric, and control treatments, we were able to test the effects certain chemicals and drugs have on the rate of paralysis, caused by the AD.

Namzaric is a pharmaceutical drug which is a combination of 28 mg memantine and 10 mg donepezil which has been used for treatment of Alzheimer's disease.[1] It offers benefits from both memantine and donepezil while allowing the patient to take fewer pills, making it the most efficient drug to use in treatment.[10] Memantine is an NMDA antagonist that has been found through past research to help decrease the effects of AD in patients by decreasing the amount of glutamate that causes neurotoxicity in the brain.[12] Excess amounts of glutamate have been thought to lead to the development of AD, and memantine so far has proven to delay the effects of AD; however, it does not appear to be a cure.[13] Donepezil is another drug which is being used to treat diseases such as dementia and AD. Donepezil is an acetylcholinesterase inhibitor, which prevents the breakdown of acetylcholine, that helps improve functions such as memory, attention, and day-to-day physical activities.[14] Studies have shown that the breakdown of acetylcholine damages the ability to store new memories; by creating a buildup of acetylcholine to block the breakdown, there is a higher stimulation of the previously disrupted receptors.[5] Studies have shown that donepezil has the potential to slow down the process and decrease the effects of AD, but cannot cure it. These drugs have only been used to treat late onset AD, and have not been tested to see whether lifelong treatment helps more in preventing symptoms.[1] We want to test how the lifelong exposure to these drugs affects the onset of paralysis of the *C. elegans*.

In previous research, coffee was shown to reduce the side effects of AD as it appears to protect against the formation of beta-amyloid plaques.[4, 7] Specifically, coffee extracts have shown to decrease the toxicity of the Aβ peptide by decreasing the onset of paralysis.[4] In the brain, coffee acts as an antagonist to the adenosine receptors which regulate tissue function. These receptors deal directly with inflammatory systems and have now become the target for many therapies in degenerative diseases such as AD.[9]

By infusing NGM plates with concentrations of both coffee and Namzaric, we compared each of the treatment plates to a control plate without any substance on it in order to determine if coffee and Namzaric have similar effects (if any) on the rate of paralysis caused by the AD gene. In creating synchronized populations and counting the amount of non-paralyzed, paralyzed, and dead worms at the same time for 12 days, we were able to use this data in a two proportion z-test to determine if the difference between our plates was significantly different, meaning that the treatments were most likely the cause of the change in paralysis.

METHODS

Through these experiments, we tested how Namzaric and coffee affect beta-amyloid plaque formations and paralysis in *Caenorhabditis elegans* (*C. elegans*), strain CL2006. These *C. elegans* are transformed with the Aβ transgene pCL12 provided by Dr. Christopher Link at CU Boulder. We used this specific strain to test how these two substances affect toxicity associated with Alzheimer's disease (AD).

Preparing Solutions

We prepared the coffee solution used in our research by following protocols originally used in the Pallanck Lab.[11] We added 18.4g of Starbucks House Blend caffeinated ground coffee beans into 100 ml of distilled water and boiled it for 30 minutes to make a concentration of 0.184g/mL (**Pic. 1**).

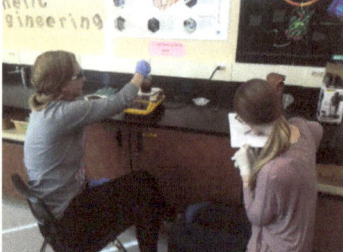

Picture 1: Lexi and Taylor preparing the coffee solution used throughout our research by boiling 18.4g of Starbucks House Blend Coffee with 100mL of distilled water for 30 minutes.

Filter paper was then used to remove all unwanted particles from the solution (**Pic. 2**). This solution was autoclaved at 250 °F for 30 minutes before being stored at 4°C in a sealed flask.

Picture 2: Lexi and Taylor filtering the coffee solution to remove the insoluble particles suspended in the solution.

Our Namzaric solution was prepared by opening up the enteric coating of 18 Namzaric pills, each containing 28 mg memantine and 10 mg donepezil, and removing the medication (**Pic. 3**). Using a mortar and pestle, the Namzaric pill contents were ground up to break the capsules of the slow-release memantine in order to completely dissolve the drug in solution. The ground-up contents were then mixed with 18 mL of sterile DI water and 100 mL of 95% ethyl acid until all of the memantine and donepezil were

Picture 3: The Namzaric pills and mortar and pestle used to ground the Namzaric medication to mix with the distilled water and ethanol to create the Namzaric solution which was fractionally distilled.

dissolved (**Pic. 4**). Memantine has a water solubility of 35 mg/mL and donepezil has an extremely low solubility in water at 2.93 mg/L; however, it has a solubility of 2 mg/mL in ethyl acid (ethanol).[3,8] Therefore, we had to create a solution using ethanol and water, and later evaporate the ethanol using fractional distillation, leaving us with the supersaturated solution of water and Namzaric. In order to remove the insoluble ingredients in the Namzaric pills, this solution was microcentrifuged to pellet the insoluble particles before removing the homogeneous solution of water, ethanol, memantine, and donepezil (**Pic. 5**). This solution was then stored at 4°C in a sealed flask.

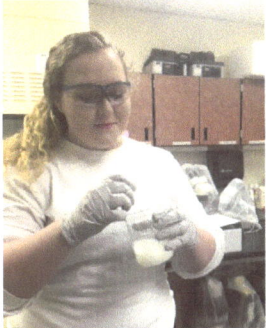

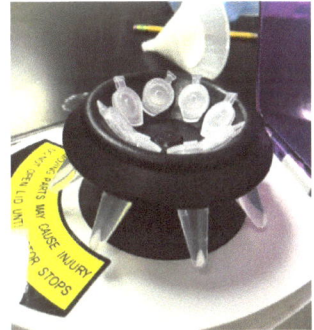

Picture 4: Allie mixing the Namzaric powder with the water and ethanol solution before performing fractional distillation.

Picture 5: Demonstrating the pellets of insoluble particles formed after microcentrifugation of the Namzaric solution to remove the insoluble particles.

Preparing NGM Plates

For both the dose determination trial and official trials completed in this research, we used 60mm Nematode Growth Media (NGM) agar plates, purchased from IPM Scientific. When preparing the plates for testing, we followed the same procedure throughout the experiment. 1mL (dose determination trials) or 750µL (official trials) of coffee, Namzaric, and control solutions were evenly distributed onto the surface of the NGM agar (**Pic. 6**) and were allowed to diffuse into the media for 24 hours at room temperature (**Pic. 7**). Then, in a biological safety cabinet, we placed two drops of an overnight culture of *Escherichia coli* (*E. coli*) strain OP50 culture to seed them before using an L-spreader to evenly distribute the culture over the surface of the plate (**Pic. 8**). These plates were allowed to sit for another 24 hours at room temperature in the biological safety cabinets before any adult worms were placed onto the surface of the plates. The plates not currently being used were stored at 4°C.

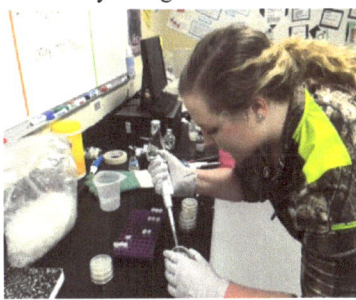

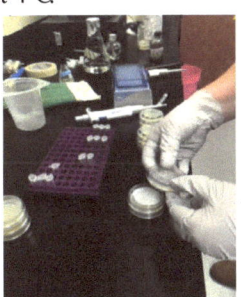

Picture 6: A picture of Allie adding the experimental coffee and Namzaric solutions onto the surface of the NGM plates for use in the dose determination trials and official trials.

Picture 7: A photo of Allie swirling the solution all around the plate to evenly disperse the solution so it can evenly diffuse into the NGM agar.

Dose Determination Trials

We first completed a dosage determination trial with both Namzaric and Starbucks House Blend coffee to determine the concentration of each substance that would be best to use in our official trials. We tested four different concentrations of Namzaric at 100% (38mg/mL), 50% (19mg/mL), 25% (9.5mg/mL), and 12.5% (4.75mg/mL). In addition, we tested two different concentrations of coffee at 100% (.184g/mL) and 50% (.092g/mL). We supplemented these plates with 1ml of each solution (**Pic. 9**) and micropipetted off the excess which did not diffuse. We also prepared control treatment plates along with the experimental treatment plates, which contained no drugs or chemicals but sterile water, to compare the rate of paralysis and stages of development in order to determine which concentrations of coffee and Namzaric would be best to use in our official trials. We created synchronized populations of *C.elegans* on each treatment plate by placing adult worms on the agar for 2 hours and then removing them, leaving only the eggs they laid on the plate. By observing the sample synchronized population worms over a 10 day period, we determined an optimal experimental Namzaric concentration of 6% (2.375mg/mL) and a 100% (0.184g/mL) concentration of the coffee solution for our official trials, as these concentrations appeared to be non-toxic and most effective towards the reduction of paralysis.

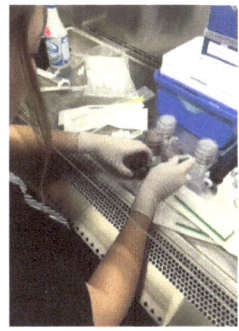

Picture 8: Brooke spreading the OP50 culture solution over the surface of the NGM plates using an L-spreader after two drops of the OP50 culture were placed on the surface of the plate.

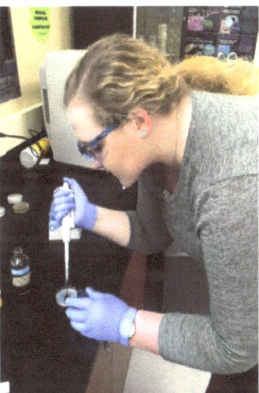

Picture 9: A picture of Allie diffusing the coffee and Namzaric solutions onto the NGM plates used in the dose determination trials before allowing the solution to diffuse into the media for 24 hours.

Official Trials

In our official trials, we completed three different treatments: Namzaric, coffee, and a control. In order to test the effect Namzaric and coffee have on the rate of paralysis, we diffused the prepared solutions into the NGM agar. We prepared six petri plates total for each treatment, all of which were prepared with the same concentration of solution. Each plate was given 750µL of its solution. After creating a synchronized population using adult *C. elegans* actively laying eggs (**Pic. 10**), these plates were stored in a incufridge at 16°C.

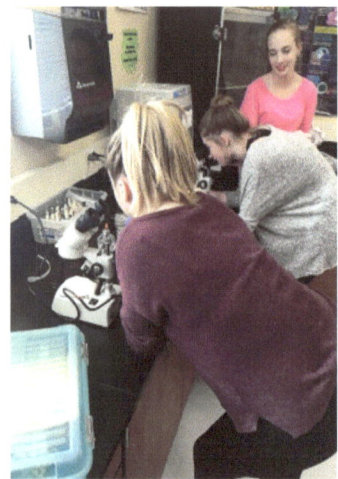

Picture 10: A photo of Brooke and Taylor picking adult *C. elegans* onto the surface of our plates to allow them to lay eggs to create a synchronized population to study over the 12 days of data collection.

Seven days after the eggs were laid, when the synchronized populations reached adulthood, we isolated the synchronized populations by moving these select worms onto another petri plate with the same treatment to allow them to continue to progress through their lifespan. For the next 5 days, we came in to count and record the number of non-paralyzed worms, paralyzed worms, and dead worms at approximately the same time each day. We considered non-paralyzed, alive worms to be any worm with locomotion in its center; paralyzed worms to be any worm with no motion in its center, but movement around the head; and dead worms to be any worm that showed no signs of movement in its center or head. We observed these worms for a total of 12 days.

RESULTS

We conducted multiple statistical tests to compare the rate of paralysis between the three treatments. In order to conduct these tests, we counted the number of paralyzed worms (**Pic. 11**) each day of our official trials out of the total number of worms in the synchronized populations created on each treatment plate and recorded the data for 12 days (**Table 1**).

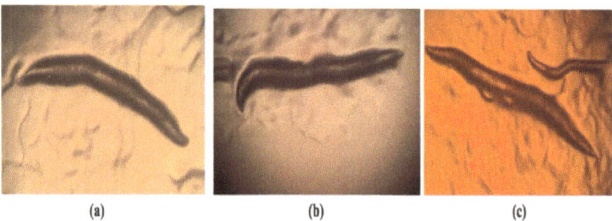

Picture 11: Paralysis of CL2006 adult *C. elegans* on the control (**a**), Namzaric (**b**), and coffee (**c**) treatment plates.

To create a visual representation of how the Namzaric and coffee slowed down the rate of paralysis, we created an ogive (**Graph 1**) using percentages of paralysis over the five days of adulthood for the control, coffee, and Namzaric treatments (**Table 1**). The CL2006 *C. elegans* reached adulthood on the 7th day of observing the synchronized populations, denoted as Day 0 of adulthood. Our calculated margin of error for each treatment was ±13.19% for control, ±10.01% for coffee, and ±10.64% for Namzaric, demonstrated by the standard error bars in **Graph 1**. This demonstrated that we are 95% confident the true average proportion of worms paralyzed each day on each plate is our estimated amount plus or minus its margin of error.

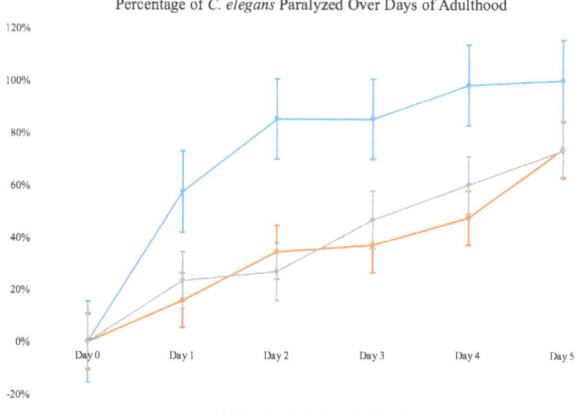

Percentage of *C. elegans* Paralyzed Over Days of Adulthood

Graph 1: This ogive graph demonstrates the percentages of *C. elegans* paralyzed over the five days of adulthood for all control, coffee, and Namzaric treatments. This graph also contains the calculated standard error bars for each treatment.

To test statistical significance, we completed a two proportion z-test over all five days of adulthood to calculate z-scores and a 95% confidence interval for our combined data each day. From there, we also calculated the probability (p-value) of randomly getting our results to ensure that the Namzaric and coffee truly were causing the change and our results weren't just due to randomness. We completed these tests to compare the control treatment with the coffee treatment (**Table 2**), the control treatment with the Namzaric treatment (**Table 3**), and the coffee treatment with the Namzaric treatment (**Table 4**).

Proportions of *C. elegans* Paralyzed Over Days of Adulthood

Plate	Day 0		Day 1		Day 2		Day 3		Day 4		Day 5	
	Fraction of Paralysis	Percentage Paralyzed	Fraction of Paralysis	Percentage Paralyzed	Fraction of Paralysis	Percentage Paralyzed	Fraction of Paralysis	Percentage Paralyzed	Fraction of Paralysis	Percentage Paralyzed	Fraction of Paralysis	Percentage Paralyzed
Control	0/54	0%	31/54	57.4%	46/54	85.2%	46/54	85.2%	53/54	98.1%	54/54	100%
Coffee	0/38	0%	6/38	15.8%	13/38	34.2%	14/38	36.8%	18/38	47.4%	28/38	73.6%
Namzaric	0/38	0%	7/30	23.3%	8/30	26.7%	14/30	46.7%	18/30	60.0%	22/30	73.3%

Table 1: This data table contains the number of worms in each treatment paralyzed per the total number of worms in the synchronized population as well as percentage of paralysis over the 5 days of adulthood which data was collected.

Statistical Significance of Variance Between Control and Coffee Treatments

	Day 1	Day 2	Day 3	Day 4	Day 5
z-score	4.000	5.020	4.794	5.714	3.993
$\hat{p}$	0.402	0.641	0.652	0.772	0.8913
p-value	<0.0001	<0.0001	<0.0001	<0.0001	<0.0001

Table 2: This data table contains the z-scores, $\hat{p}$ (sample proportion) scores, and p-values calculated through the two proportion z-test between the control and coffee treatments over the last 5 days of adulthood.

Statistical Significance of Variance Between Control and Namzaric Treatments

	Day 1	Day 2	Day 3	Day 4	Day 5
z-score	3.006	5.360	3.740	4.630	3.989
$\hat{p}$	0.4523	0.6429	0.714	0.845	0.9048
p-value	=0.001	<0.0001	<0.0001	<0.0001	<0.0001

Table 3: This data table contains the z-scores, $\hat{p}$ (sample proportion) scores, and p-values calculated through the two proportion z-test between the control and Namzaric treatments over the last 5 days of adulthood.

Statistical Significance of Variance Between Coffee and Namzaric Treatments

	Day 1	Day 2	Day 3	Day 4	Day 5
z-score	0.7855	-0.6686	0.8174	1.036	-0.0326
$\hat{p}$	0.1912	0.3088	0.4118	0.5294	0.7353
p-value	=0.284	=0.248	=0.293	=0.350	=0.013

Table 4: This data table contains the z-scores, $\hat{p}$ (sample proportion) scores, and p-values calculated through the two proportion z-test between the coffee and Namzaric treatments over the last 5 days of adulthood.

All of the z-scores calculated were above 3, some ranging as high as 5.514, and all of the calculated p-values less than or equal to 0.0001 when the experimental coffee and Namzaric treatments were compared to the control treatment.

In **Tables 2-4**, we recorded the data from conducting our two proportion z-tests, where the z-score is the standard deviations between the two treatments, $\hat{p}$ is the total proportion of worms paralyzed on both plates combined, and p-value is the probability of getting our results randomly for the data collection on each day. If the z-score is above 3, it is considered statistically different. Because all z-scores in **Tables 2-3** are above 3, the likelihood of the results being randomly are statistically unlikely. If there is a higher z-score, the p-value will be lower because it is less likely to get that result. The lower the probability, the less likely it is we got our data due to randomness, and the Namzaric and coffee actually affected the rate of paralysis. To test the statistical variance between the coffee and Namzaric treatments, we ran the same two proportion z-test to compare the two treatments to each other. As the calculated z-scores are not greater than 1.1 or lower than -1.0 and all the p-values are greater than 0.01, this demonstrates that there isn't a statistically significant difference between the two experimental treatments, showing that both Namzaric and coffee both had similar effects on the rate of paralysis in *C. elegans* strain CL2006.

QR Code: To view the CL2006 *C. elegans*, scan this QR code and watch the short video we created.

Paralysis, as shown through the margin of error bars (**Graph 1**), z-scores, $\hat{p}$-scores, and p-values (**Tables 2-4**), paralysis is statistically different between the control and the two experimental treatments. Furthermore, there is no statistical difference between the two experimental treatments on paralysis.

DISCUSSION

Being the sixth leading cause of death in the United States, Alzheimer's Disease (AD) is a destructive neurodegenerative disease that affects millions of people all around the world. Through the study of how coffee and Namzaric affect the toxicity of AD, we were able to determine if these two experimental treatments had a statistically significant difference on the rate of paralysis in comparison to a control treatment, which was exposed to no drugs or other substances. This was completed through analyzing the rate of paralysis in *Caenorhabditis elegans (C. elegans)* strain CL2006 when exposed to these various treatments over the lifespan of 12 days. Our original hypothesis was that both the coffee and Namzaric would decrease the rate of paralysis in the *C. elegans* strain CL2006; however, we did not know if one would work better than the other or if they would both have the same effect. From our statistical analysis, we were able to conclude that both coffee and Namzaric have a statistically significant effect in decreasing the rate of paralysis in this particular strain of *C. elegans* in comparison to a control treatment. In addition, we were able to determine that the effects of coffee and Namzaric treatments on paralysis, when compared to each other, were not significantly different, suggesting that both have a statistically significant effect on decreasing the rate of paralysis in *C. elegans* due to Aβ plaque formation.

Through analyzing the ogive graph (**Graph 1**), it is apparent that the control treatment *C. elegans* had a much steeper incline in the number of paralyzed adults over the days of adulthood in comparison to the coffee and Namzaric treatments. It's also relevant that the Namzaric and coffee slopes are very similar, meaning they both similar effects in their relation to the rate of paralysis.

Although, in order to identify that the coffee and Namzaric had statistically significant effects on the rate of paralysis, we completed two proportion z-tests on the data to determine the z-scores, and consequently the p-values, as well as the $\hat{p}$-scores to determine the significance in our data. When the control was compared to both the coffee and Namzaric, the z-scores were all very high, ranging from 4.000 to 5.714 (**Table 2**) for coffee and 3.006 to 5.360 (**Table 3**) for Namzaric over the five days of adulthood (days 1-5). Because of these scores, we can conclude that both coffee and Namzaric had a statistically significant difference on the onset of paralysis in these *C. elegans*. In addition, because all of the probabilities (p-values) are so low, we can conclude that the coffee and Namzaric are most likely causing the change in rate of paralysis, and that our results are not simply due to randomness.

When we completed the two proportion z-test between coffee and Namzaric treatments, because of the low z-scores calculated (**Table 4**), this demonstrates that there is no statistical difference between the data of the two treatments, as the z-scores ranged from -0.6686 to 1.036, meaning they both had similar effects on the rate of paralysis, and one did not statistically decrease the rate of paralysis significantly more than the other.

Furthermore, to demonstrate that there is a statistical difference in the data between treatments another way, we calculated the margin of error bars for the control, coffee and Namzaric plates to create a 95% confidence interval, we determined them to be ±13.19% for control, ±10.01% for coffee, and ±10.64% for Namzaric. When these were graphed, (**Graph 1**), the margin of error bars for the control did not overlap with either the coffee or Namzaric margin of error bars beyond Day 0, when the worms hit adulthood. Meaning that even if we were at the top of our interval for control and the bottom of our interval for either treatments, they still wouldn't be close to each other, proving that there is a statistically significant difference. On the other hand, the coffee and Namzaric standard error bars overlapped all 5 days of adulthood, demonstrating that they can not be considered statistically significantly different.

In comparison to the research conducted by Lubin and Link and Dostal on the effects of Starbucks House Blend Coffee (caffeinated) on *C. elegans* strains CL4176, CL6176, CL6180, CL6222, and CL2337, coffee was also determined to have a significant effect on the rate of paralysis.[4,7] In this research, as the Aβ was inducible in these strains, by 26 hours after induction, all of the control treatment population were 100% paralyzed, and at 30 hours after induction, 60% of the population treated with coffee were paralyzed. We found similar results in our research with the CL2006 strain *C. elegans* as on the fifth day of adulthood, 100% of the control treatment were paralyzed while 73.6% were paralyzed in the coffee treatment.

From these results, we are able to determine that both coffee and Namzaric have statistically significant results in slowing the rate of paralysis in *C. elegans* strain CL2006, transformed with the human Alzheimer's disease gene to produce Aβ plaques. In addition, we are able to conclude that there is no statistical difference between the coffee and Namzaric treatments, suggesting that coffee has a similar effect on decreasing the rate of paralysis as Namzaric, an expensive pharmaceutical drug used to treat Alzheimer's disease in humans, when the *C. elegans* are exposed to the treatments over their entire lifetime. These results show that in this particular strain of *C. elegans*, Namzaric has no statistically significant difference in decreasing the rate of paralysis, caused by AD, in comparison to coffee. From these results, we conclude that Starbucks House Blend coffee works just as efficiently as Namzaric, an expensive

pharmaceutical drug, in slowing the toxicity of Alzheimer's disease in *C. elegans* strain CL2006.

Throughout our experiment, there were a variety of possible sources of error which may have influenced our results. One of the largest sources of error was due to the small population sizes we collected data from. Because our sample sizes ranged from 30 to 54 organisms, the population sizes are relatively small, meaning that this data may not be accurate enough to project onto the rest of the CL2006 populations alive. In addition, there were inconsistencies between the sample sizes between the control and experimental treatments, also a possible source of error in calculating our statistics. Moreover, half of the coffee data and half of the control data were collected at different times due to errors in technique, so at one time of day more or less worms could be paralyzed compared to another time. This could have also greatly impacted our collected data and results by limiting our ability to compare when scoring the *C. elegans*. Throughout this experiment, there were four different people collecting and counting data for paralysis over the twelve days of our official trials, so it's possible that a worm one would have counted as paralyzed, was considered dead/non-paralyzed to another. When determining the dosage of Namzaric to use in the official trials, we ran into issues with the pH of the Namzaric solution. When the solution was diffused onto the NGM plates, seeded, and then populated, the growth of the *C. elegans* was stunted. The ethanol in the Namzaric solution made it acidic, which required us to use a very low dose of Namzaric in our official trials. As a result, from looking at the *C. elegans* under a microscope for an extended period of time, the *C. elegans* may have been exposed to heat for long periods of time, which harms the worms and could possibly even kill them. This means that perhaps not all paralysis/death counted daily was a result of the Aβ plaque formations. Overall, these possible sources of error could have influenced the paralysis of the *C. elegans* in the synchronized populations, altering our collected data and subsequently the statistical analysis completed. Moreover, these sources of errors could affect the accuracy of our results.

There are various next steps for this research. First, the effects of decaffeinated coffee versus caffeinated coffee should be studied to determine the role caffeine plays on the rate of paralysis. In addition, higher doses of Namzaric should also be studied to see if that has a more effective impact on paralysis. Coffee and Namzaric should also be studied for their effects on how they impact paralysis when the *C. elegans* are only exposed to the treatments once they reach adulthood to determine if these substances have any significant effects on paralysis in the later stages of life, as Alzheimer's disease is a neurodegenerative disease that takes effect later on in the human lifespan.

ACKNOWLEDGMENTS
First, we would like to thank Dr. Christopher Link of University of Colorado, Boulder for mentoring our research, providing us with the *C. elegans* needed, and aiding us in the knowledge needed to further our research and decide on a research question.

We are thankful for Robin Fordham, PA, who provided us with the Namzaric we used throughout our research. We would also like to thank Amy Hacker and Susanne Petri for sharing their lab space with us, supporting our research, and helping us come in on weekends to further our research; Rock Canyon High School for providing laboratory space and equipment; and Douglas County School District for the Innovation and Perkins Grant funding that provided research grade laboratory equipment. We would like to thank Tom Dillon of the Community College of Aurora for his feedback and support with the design and implementation of our project. We would also like to give a huge thank you to everyone who gave donations to help support and fund our research, every donation, big and small, helped in allowing us to gather our materials. A huge thank you to David Ferguson for providing us with multiple materials necessary to complete our experiment, for aiding us in our process of creating our solution of Namzaric, and for being a consistent support to a class that needs him and values his expertise; and Bryan Winkelman, for helping us in finding resources to gather materials and knowledge from as well as for our website, course design, and help in creating the final published journal article. Finally, a big thank you to Matt Gracey and Dr. Jason Dunkle for aiding us in statistically analyzing our data.

REFERENCES

1. Boinpally, R., Chen, L., Zukin, S., McClure, N., Hofbauer, R., & Periclou, A. (2015, May 28). A Novel Once-Daily Fixed-Dose Combination of Memantine Extended Release and Donepezil for the Treatment of Moderate to Severe Alzheimer's Disease: Two Phase I Studies in Healthy Volunteers. *Clinical Drug Investigation*, 35(7), 427-35. Doi: 10.1007/s40261-015-0296-4. Retrieved 2015, October 7. [Web]

2. Corsi, A., Wightman, B., & Chalfie, M.. A Transparent Window into Biology: A Primer on Caenorhabditis Elegans. *WormBook* (2015): 1-31. 18 June 2015. Retrieved 2015, September 24. [Web]

3. Donepezil. National Center for Biotechnology Information. (2015, December 6). PubChem Compound Database. Retrieved 2016, January 13. [Web]

4. Dostal, V., Roberts, C., & Link, C. (2014, March 17). Genetic Mechanisms of Coffee Extract Protection in a Caenorhabditis elegans Model of -Amyloid Peptide Toxicity. *Applied and Environmental Microbiology*, 78(7), 2075–2081. Doi: 10.1128/AEM.07486-11. Retrieved 2015, October 7. [Web]

5. Hasselmo, M. (2006, September 29). The Role of Acetylcholine in Learning and Memory. *Current Opinion in Neurobiology*, 16(6), 710-715. Doi: 10.1016/j.conb.2006.09.002. Retrieved 2015, October 7. [Web]

6. Latest Alzheimer's Facts and Figures. Alzheimer's Association. *Latest Facts & Figures Report*. (2013, September. 17) Alzheimer's Association. Retrieved 2015, September 24. [Web]

7. Lublin, A. & Link, C. (2012, Mar. 10) Alzheimer's Disease Drug Discovery: In-vivo Screening Using C. Elegans as a Model for β-amyloid Peptide-induced Toxicity. *Drug Discovery Today. Technologies.*, 10(1), e115-e119. U.S. National Library of Medicine. Retrieved 2015, September 24 [Web]

8. Memantine. National Center for Biotechnology Information. (2015, December 6). PubChem Compound Database. Retrieved 2016, January 13. [Web]

9. Nehlig, A., Daval, J., et al (1992, August 17). Effects of coffee/caffeine on brain health and disease: What should I tell my patients? *Practical neurology*, 16(2), 89-95. Doi: 10.1136/practneurol-2015-001162. Retrieved 2015, October 7. [Web]

10. PR Newswire. (2015, May 18). Actavis Launches NAMZARIC™ (memantine hydrochloride extended-release and donepezil hydrochloride), a Fixed-Dose Combination Therapy for the Treatment of Moderate to Severe Alzheimer's Disease. Retrieved 2015, October 4. [Web]

11. Trinh, K., Andrews, L., Krause, J., Hanak, T., Lee, D., Gelb, M., & Pallanck, L. (2010, April 21). Decaffeinated Coffee and Nicotine-Free Tobacco Provide Neuroprotection in Drosophila Models of Parkinson's Disease through an NRF2- Dependent Mechanism. *The Journal of Neuroscience*, 30(16), 5525-5532. Doi: 10.1523/JNEUROSCI.4777-09.2010. Retrieved 2015, October 7. [Web]

12. U.S National Library of Medicine. (2015, August 25). Memantine: MedlinePlus Drug Information. Retrieved 2015, October 7. [Web]

13. U.S. National Library of Medicine. (2013, July 18). Alzheimer's disease: Does memantine help? Retrieved 2015, October 7. [Web]

14. U.S. National Library of Medicine. (2015, August 25). Donepezil: MedlinePlus Drug Information. Retrieved 2015, October 7. [Web]

ABOUT THE AUTHORS

Pictured: Brooke (left) and Allie (right). Not pictured, our mentor, Dr. Christopher Link with the University of Colorado, Boulder.

My (Allie) interest in Alzheimer's Disease was sparked when I began working at a nursing home in the Memory Care Unit. From this experience, I found a passion in wanting to research neurodegenerative diseases that cause dementia in the later stages of life. Ultimately, this is what drove the research we completed this year. After high school, I plan on majoring in biochemistry with a focus on neuroscience and then attending medical school and becoming a medical doctor and earning my PhD with a focus on neurodegenerative diseases.

I (Brooke) pride myself in not only being adept in science in math, but with English and the arts as well. I've thoroughly enjoyed learning more about Alzheimer's and how chemicals function in the brain, and, even more so, loved getting to test my own innovative thoughts and ideas in a supportive and caring environment. In college I plan to further my interest in Biotech and work in a research lab; while at the same time pursuing my passion for film and music, and learning to combine science with art.

Through this research experience, we have learned how to be a leader and work with a team and how to conduct research in a lab with peers. In addition, this class taught us the essential knowledge, collaboration skills, and problem solving skills required when conducting real scientific research. This outstanding research opportunity has vastly opened up our knowledge and capabilities as students and researchers, providing us with one of the most valuable experiences of our high school career. So many doors have been opened for our future and we are very thankful for this experience as it has shaped the course of our lives.

The effects of CRISPR folding on the efficiency of targeting the lacZ gene site within *E. Coli* using a blue/white screen

Eugene Kim, William E. Gibbons, Dr. Sarah M. Richardson, and Shawndra L Fordham

Department of Science, Principles of Experimental Design in Biotechnology, Rock Canyon High School, Highlands Ranch, Colorado, USA

This study examines the CRISPR gene editing system, typically found in bacteria, and how folding or orientation of its spacers can affect the efficiency of the system when inserted into another organism. Using *Escherichia coli* as the organism for this study, three trials of transformations testing 17 plasmids were performed with each examining the effect of three features that affect the secondary structure of CRISPR (folding, orientation, and length) in order to see if they affect the gene-editing efficiency. The efficiency was predicted using the software package toaster, and then analyzed using a blue-white screen on X-gal plates, with transformed colonies appearing blue, and untransformed white. Calculating transformation efficiency allowed us to determine if there was truly a difference produced by changing the shape of the CRISPR complex. Eight of the 17 plasmids matched the efficiency predicted by toaster. In comparison to one another, using standard error, our data produced results that indicate that the secondary structure of CRISPR does affect the efficiency of the system. However, despite this data, no concrete conclusions can be drawn because of our inaccurate control. While our control should have had near 0% efficiency, we had 51.78% efficiency instead. This is an indication that our other data is inaccurate as well. Ideally, these findings will contribute to the growing body of knowledge surrounding CRISPR technology that will ultimately improve the efficiency of CRISPR systems used, allowing them to be used as a more permanent solution to gene-editing in comparison to systems such as RNA interference.

CRISPRs (Clustered Regularly Interspaced Short Palindromic Repeats) are starting to become a very important topic in the biotechnology field as an accessible gene-editing technology. Editing genes utilizing CRISPR technology has been observed to be more flexible, and most importantly, much cheaper, than traditional gene-editing technologies. A traditional gene-editing technology like zinc finger nuclease technology presently costs at least $5000, compared with the $30 usually spent for CRISPR.[4] As more information about CRISPR technology becomes uncovered through research, the amount of applications of the technology increases. For example, in a paper published in 2015, it was reported that the CRISPR-Cas9 system was utilized to study the deadly pathological yeast, *Candida albicans*.[11]

The most exciting aspect of this technology is in its potential. Other existing applications include advances in bioengineered animals and plants, such as beagles with double the amount of muscle and mildew resistant wheat.[8,12] Scientists are working on creating chickens with non-allergenic eggs, healthier bees, malaria-resistant mosquitoes, and plans to bring extinct animals back to life by changing the genome of their relatives.[8] The possibilities are endless in the field of CRISPR technology, and fortunately, many are quick to harness these possibilities.

However, for all its advantages and potentials, an adolescent technology is never perfect. For CRISPRs, the low efficiency in eukaryotes needs to be improved before it can be widely applied.[10] In nature, CRISPR acts as a defensive mechanism against foreign DNA, largely viral DNA and plasmid DNA in bacteria, so it is no surprise that the efficiency in bacteria is very high.[6] In eukaryotes, however, the efficiency decreases significantly. In this investigation, we strove to further understand the possibility of improving the overall efficiency of CRISPRs through manipulating its secondary structure.

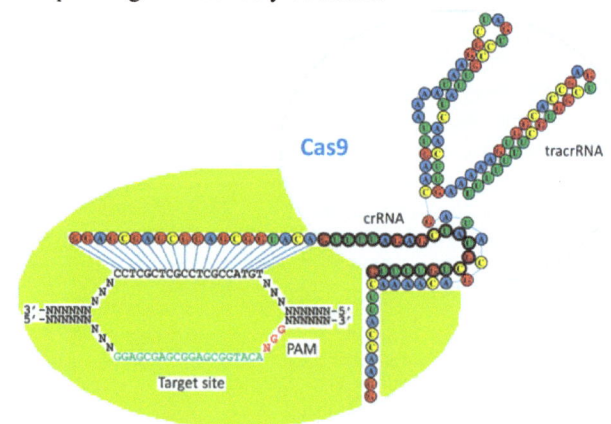

Figure 1: This is a diagram of a CRISPR-Cas9 complex.[3]

To understand why manipulating secondary structure might improve efficiency, it is critical to understand how CRISPRs work. Currently, there are three types of CRISPR systems that have been identified in nature. Type 1, found both in bacteria and archaea, is defined by the Cas3 protein found in all type 1 CRISPRs, type 2, found naturally solely in bacteria, is defined by the Cas9 protein, and type 3, usually found in archaea, but also in some bacteria, is defined by the Cas10 protein and Cas6.[2,7,9] Cas3, Cas9, Cas10, and Cas6 are called "signature proteins."[1] Of the three systems, the type 2 CRISPR system is the simplest and the most researched, and thus, the most commonly used. In the type 2 CRISPR system, a combination of small RNA spacers that makes up the CRISPR RNA (crRNA), forms a complex with a trans-activating CRISPR RNA (tracrRNA). Spacers are bits of viral DNA (called protospacers) that exist in the CRISPR sequence in the form of a viral memory.[1] Since tracrRNA is complementary to crRNA, it binds by base pairings and is cleaved with RNase III to make the Guide RNA (gRNA). The gRNA then binds to the Cas9 protein. This combination is called the CRISPR-Cas9 complex **(Fig. 1)**. The combination then locates a place on the DNA that matches to the target sequence dictated by the gRNA. The section to cut is recognized by two things: the protospacers that have the corresponding spacer in the gRNA and sequences of DNA called Protospacer Adjacent Motif (PAM) that flank the protospacer. Having the PAM helps in that it prevents the CRISPR-Cas9 from cutting the spacer out of the bacteria's own genome since the bacterial spacer is adjacent to CRISPR repeats rather than the PAM.

We think that different secondary structures of the gRNA, as well as its strand orientation and length, may impact how well it binds with the Cas9 protein. Since the Cas9 needs to bind to the gRNA in order to cut the DNA and insert a new sequence, different CRISPR secondary structures may impact the efficiency of the entire system.

We predicted the efficiency of the different gRNA secondary structures by using the software package Dr. Sarah Richardson created, called toaster. Toaster calculates a score for a sequence of gRNA by predicting what a structure formed by this sequence would look like and comparing it with the "hairpin" structure that is predicted to give the highest efficiency **(Fig. 2)**. Based on this, we can predict whether a plasmid would pass (have an overall higher efficiency) or fail (have an overall lower efficiency).The hairpin structure is predicted to have the highest efficiency based on toaster because it has a section where the Cas9 can clearly recognize.

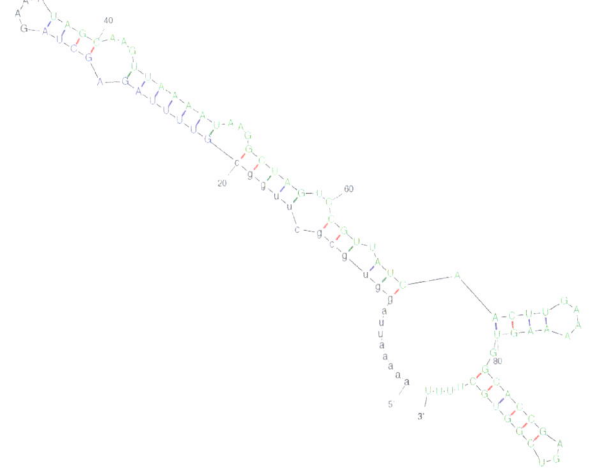

Figure 2: This (at right) is an example of the "hairpin" structure of the gRNA. It is predicted that this gRNA will have the highest efficiency.

METHODS

In this experiment, we looked at the effect of the secondary structure of the CRISPR system on its efficiency in *Escherichia coli*. In each trial 17 plasmids containing the gRNA of the CRISPR system that were transformed into *E. coli*. Successfully transformed colonies were grown on agar plates containing antibiotics, which were used as a marker to identify successfully transformed bacteria. In addition to the antibiotic marker, the plasmid had *lacZ* gene in it, which allowed the transformed colonies to uptake the X-gal in the media and consequently appear as blue. In each trial, three control plates were used, simulating the *E. coli* growth without a target site for the CRISPR Cas9 systems, in order to account for any erroneous data.

In this investigation, we strove to understand exactly what kind of secondary structures, encompassing the categories of predicted foldings, orientation of the CRISPR RNA, and length, would yield the highest efficiency. Folding refers to the different shapes of the CRISPR RNA. Orientation refers to the order in which two spacers are oriented in the CRISPR RNA, while length refers to the size of the CRISPR RNA. The first, folding, included four testing plasmids with folds predicted to yield high CRISPR efficiency and four predicted to yield low efficiency by toaster. The second, orientation, tested two plasmids with spacers in their normal PAM orientation and two with their orientations flipped. The final set, length, tested if the number of spacers affects efficiency and included two plasmids with the normal double spacers and two plasmids with only a single spacer.

We transformed a total of 17 plasmids into *Escherichia coli*, 16 plasmids listed above, and a control plasmid that did not have a fixing template to repair the DNA after ithad been broken by the Cas9 **(Pic. 1)**. All of the mentioned plasmids coded for the *lacZ,* which codes for β-galactosidase, an enzyme that cleaves lactose into glucose and galactose. Since the bacteria we used had their *lacZ* genes disrupted, we used X-gal to measure the efficiency of the CRISPRs. All plasmids were tested through the software package, toaster, to obtain predicted secondary structures and their efficiency before the experiment began.

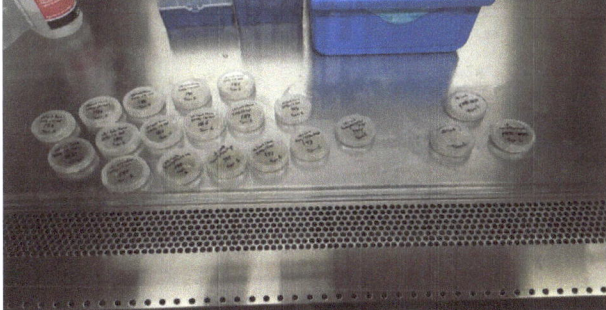

Picture 1: 20 plates prepared for the first trial within the biological safety cabinets in which they were made.

Preparation of Chemical Solutions

Before we began the experiment, several chemicals were prepared and used during transformation as well as for making the agar plates. The chemical stock solutions that were made included 1M $MgCl_2$, 5M NaCl, 1M KCl, 1M $MgSO_4$, 1 M Glucose, and 1 M $CaCl_2$. We also made a 1X

TSS Buffer solution and a 5X KCM salt solution. 60mm LB-Agar Petri plates were prepared by mixing 9mL of LB agar with 4.5uL of 100mg/mL carbenicillin, 4.5uL of 50mg/mL kanamycin, 18ug of 20mg/mL X-gal, and 9uL of 0.1M solution of IPTG per plate **(Pic. 2).**

Picture 2: The hot plate holds the LB-Agar as it is heated and stirred.

Transformation

First, a selected strain of the pSMR0027 *E. coli* bacteria, that were obtained from Dr. Sarah Richardson, was streaked onto the prepared LB plate, then incubated overnight at 37°C. Individual colonies were then cultured overnight at 37°C in lysogeny broth. It was then stored at 4°C until used. Fresh cultures were created throughout the experiment by inoculating fresh lysogeny broth with 1% of the overnight culture. This was shaken at 200 rpm at 37°C, with the goal of obtaining an OD_{600} of between 0.4 and 0.6 as measured by a spectrophotometer to obtain the absorbance units (au). The absorbance units were then converted to kletts, and the kletts to OD_{600}.

Picture 3: The overnight culture was shaken at 37.4°C at 200rpm.

The cultured cells were concentrated and resuspended with 1x TSS buffer. We then moved 100uL of cells from the TSS buffer and placed them in new microcentrifuge tubes. 1uL of 10ng/ul of the specific treatment plasmids, along with 20uL of 5X KCM was added to 100uL of the bacterial cell mixture. The volume was then increased to 200uL with distilled water. We then heat shocked the bacteria. To perform heat shock the bacterial cell mixtures were put in a hot water bath, at 42°C for 90 seconds, placed in an ice bath for three minutes, and then recovered in 500uL of SOC and shaken at 200 rpm for 60 minutes at 37.4°C to encourage cell growth **(Pic. 3).** The cells were then concentrated and decanted one last time. At this point, the remaining cells were plated onto the prepared LB-Agar plates, wrapped in tin foil (since X-gal degrades with light), and incubated at 30°C for two days, at which point the colonies were counted **(Pic. 4).**

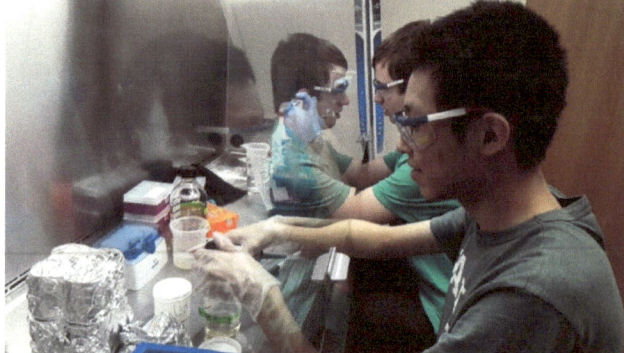

Picture 4: Will and Eugene are plating and wrapping completed plates in tin foil to avoid X-gal decomposition.

Statistics

To analyze the results, we counted the number of the white and blue colonies, made ratios of blue to total colonies, and took the average ratios of the three trials we conducted for the main data for each treatment (plasmid), shown in **Graphs 1-3** as the blue bars, and applied ±1 standard error bars. We could make reasonable conclusions for the plasmids with standard error bars that did not overlap.

RESULTS

In gathering and compiling the results, we used the ratio of blue colonies to total colonies to determine the efficiency of the plasmid, which contained the fold we were testing. So, the higher the percentage, the higher the efficiency. This ratio was used in order to more easily compare results with that of Laure M. Leynaud-Kieffer.[5]

While we made individual, relative observations, we also observed the efficiency as a whole. The basic method we used was making a criteria for "passing" and "failing", in comparison to the average transformation efficiency of the three trial data sets, which we used as the reference point. Any data point above the reference point was considered "passing"; while data below the reference point was considered "failing". These pass-fail assignments for each plasmid are shown in **Table 1**.

Table 1: Summary of Results

Plasmid Letter	DNA	Set Type	Trans. Efficiency	Toaster	Call	Other Research
Control	pSMR0027	--	51.78%	--	--	--
B	pSMR0179	Folding	23.95%	Pass	Fail	Pass
C	pSMR0180	Folding	31.60%	Pass	Fail	Pass
E	pSMR0182	Folding	75.95%	Pass	Pass	Fail
F	pSMR0183	Folding	28.70%	Fail	Fail	Fail
G	pSMR0184	Folding	45.31%	Fail	Pass	Fail
I	pSMR0186	Orientation	26.50%	Pass	Fail	Pass
J	pSMR0187	Orientation	53.07%	Pass	Pass	Pass
L	pSMR0189	Orientation	19.77%	Pass	Fail	Pass
M	pSMR0190	Orientation	58.87%	Pass	Pass	Fail
N	pSMR0191	Length	44.10%	Pass	Pass	Fail
O	pSMR0192	Length	17.20%	Fail	Fail	Fail
P	pSMR0193	Length	44.27%	Pass	Pass	Fail
Q	pSMR0194	Length	35.70%	Fail	Fail	--

Table 1: The identifying plasmid letter is the letter that we designated to allow easier categorization of the plasmids. DNA signifies the name of the plasmid given by Dr. Sarah Richardson, and the set type is the group each plasmid belongs to. The transformation efficiency is the average of our efficiency results over the three trials. The call, pass or fail, conveys whether the plasmid achieved a higher efficiency measure than the average (pass) or was lower (fail). The fifth column shows call predicted by toaster. The sixth column indicates the results of our data, while the last column indicates the results of Laure Marie Leynaud-Kieffer.' The highlighted green rows signify the agreement of our results and that of Ms. Marie Leynaud-Kieffer. It is very important to note that while the control should have had near 0% efficiency, our control had 51.78% efficiency.

We did not have enough data for plasmids A, D, H, K, and the two control plates of LB-Carb50 and LB-Carb50-Kan50 to reasonably analyze the data using standard deviation and error bars, which requires at least three trials. However, in regards to the two control plates, there were no blue colonies in either of those plates, so it was safe to assume that the white colonies to total colonies ratios were 100% for both.

Folding

Plasmids A through H were tested for folding **(Graph 1)**. Of these, we collected sufficient data for standard deviations for plasmids B, C, E, F, and G. Toaster predicted that plasmids B, C, and E would pass and that plasmids F and G would fail.

In the actual pass-fail analysis only plasmid E passed the average transformation efficiency of 40.54% as expected. Plasmid G passed when toaster predicted that it would fail.

Graph 1: This graph shows the transformation efficiency based on folding. Efficiency was calculated using the blue colonies to total colonies ratio. Higher data points mean better transformation efficiency. Plasmids E and G passed our average transformation efficiency criteria of 40.54%.

Orientation

Plasmids I, J, K, L, and M were tested to see how the direction of the spacers affected the transformation efficiency **(Graph 2)**. We could collect sufficient data for standard deviations for plasmids I, J, L, and M. Toaster predicted that all the plasmids tested for this set would pass. Plasmids J and L had flipped orientations, plasmids I and K had the normal orientations, and plasmid M did not have a fixing template. Examining the plasmids for this set's pass-fail analysis, plasmids J and M failed to pass the average criteria of 45.87%, while the control and plasmids L and I passed as predicted by toaster.

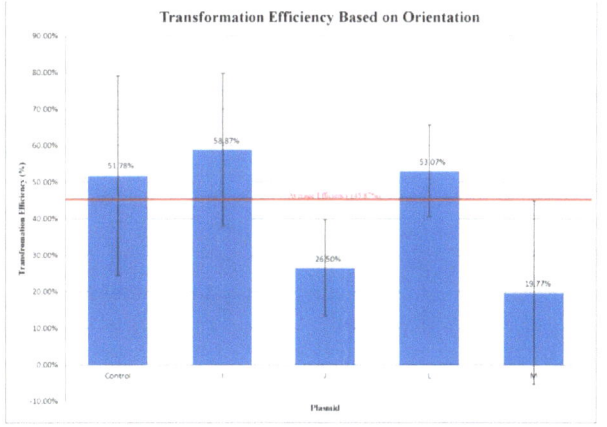

Graph 2: This graph shows the transformation efficiency based on orientation. Efficiency was calculated using the blue colonies to total colonies ratio. Higher data points mean better transformation efficiency. Plasmids L and I passed our average transformation efficiency criteria of 45.87%.

Length

Plasmids N, O, P, and Q were tested for length **(Graph 3)**. Plasmids N and O had smaller sizes, their deletion size of 870 bp, compared to the 4192 bp of plasmids P and Q. We collected sufficient data for standard deviations for all the plasmids in this set. Toaster predicted plasmids N and P would pass and plasmids O and Q would fail. In the actual pass-fail analysis, toaster correctly predicted the efficiency for all the plasmids in the set. The control, plasmids N and P passed, and plasmids O and Q failed the average criteria of 38.73%.

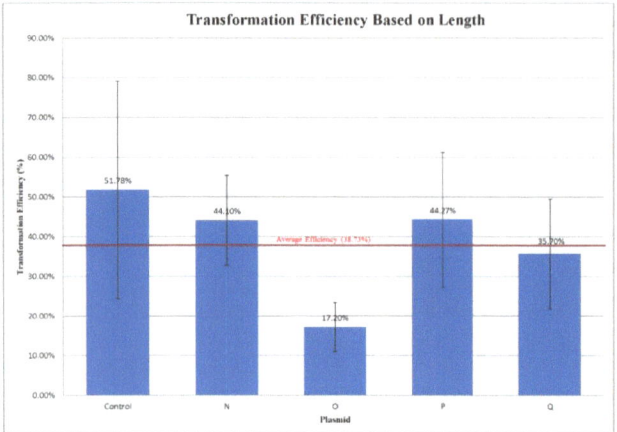

Graph 3: This graph shows the transformation efficiency based on length. Efficiency was calculated using the blue colonies to total colonies ratio. Higher data points mean better transformation efficiency. Plasmids N and P passed our average transformation efficiency criteria of 38.73%.

DISCUSSION

Our results were created on the basis of passing, having a percentage of blue colonies to total colonies on the plate higher than the average transformation efficiency; or failing, in which case the percentage of blue colonies to total colonies on the plate is less than the average transformation efficiency. By comparing the standard deviation and errors of each trial we were able to determine that the differences in the folding, orientation, and length of the CRISPR sequences creates statistically significant differences in the efficiency of the CRISPR. However, due to our small sample size, a test could not be conducted to determine that the results were not due to random chance.

We compared our data to the calls predicted by toaster. Overall, our data matched with the toaster predictions for eight plasmids. Of those eight plasmids, we had two matches in the folding set (plasmids E and F), two matches in the orientation set (plasmids J and M), and four matches in the length set (plasmids N, O, P, Q).

Our findings were also compared to the findings of Laure M. Leynaud-Kieffer.[5] While we used averages to determine our pass fail, she used the Q-test. Nearly all of our calls for the plasmids conflicted with her findings, except for one plasmid in each set: plasmid F (folding); plasmid J (orientation); plasmid O (length) **(Table 1)**. In these plasmids our results matched that of Ms. Leynaud-Kieffer.

The agreement of data and prediction for these plasmids suggest that there are differences in efficiency of CRISPR based on changes made to their form, such as their folding, their orientation, or their length. These changes in efficiency in comparison to one another could be attributed to a few different factors: how similar the CRISPR is to the original sequence that is being edited within the *E. coli* or possibly different secondary structures of the CRISPR binding to the site with varying success in transformed cultures due to the base pairing and shape that the CRISPR takes on. Furthermore, the data may be skewed due to

overgrowth on the plates, specifically some plates in the second and third trial growing lawns of bacteria. This led to the inability to count colonies on these plates; in our case, these overgrown plates were plasmid K (pSMR0188) in trial two, and plasmids A (pSMR0178), D (pSMR0181), and H (pSMR0185) in trial three. Plasmids A, D, and H were tested for folding, and plasmid K was tested for orientation.

We cannot make conclusions related to efficiency for most of these plasmids as of yet, because of differences between our calls, and the calls of Ms. Leynaud-Kieffer. It is interesting to note that for the calls where both our research agreed, plasmids F, J, and O, the toaster prediction agreed with those calls as well. This suggests that our own data does add to the growing body of knowledge related to CRISPR efficiency and the ability of toaster to predict efficiency.

It is very important to note, however, that our control plasmid had an efficiency of 51.78%. This is critical because this control plasmid should have had an efficiency of almost zero, as the control lacked the ability to recognize the lacZ gene site and should therefore be extremely unlikely that it will find, bind to, and edit that gene site. Because our control differs so much with what it should have been, we cannot form any conclusions about these results, as the data gathered may have been completely skewed , as signified by the incorrect control data generated by the experiment.

Conclusion

While the results from our experiments and the experiment conducted by the other researchers suggest that certain secondary structures of the CRISPR in our sets of folding, orientation, and length may indeed affect efficiency, because our control was compromised, we cannot form any significant conclusions from the data we have collected. There is certainly more work to be done regarding this experiment. We lacked enough data for some of the plasmids, namely plasmids A, D, H, and K, and our control plate was inaccurate. Furthermore, more trials should be conducted to verify the data with a more accurate standard deviation. For future reference, it should also be noted that the bacteria are not to be grown for longer than 48-72 hours; allowing them to ceaselessly multiply will result in the possible inability to count bacterial colonies in that plate. Regarding the idea that secondary structure affects efficiency, further experiments in testing other possible structural types other than folding, orientation, and length, and why certain structures cause a decrease or increase in efficiency will be essential in deepening understanding in this field.

ACKNOWLEDGMENTS
We would like to offer our gratitude to all those who have helped contribute to this project. Furthermore, we would like to thank Amy Hacker and Susanne Petri for sharing their laboratory space, Rock Canyon High School for providing all the necessary equipment and the laboratory equipment, and Douglas County School District for the Innovation and Perkins Grant funding that provided research grade laboratory equipment. We offer our

thanks to David Ferguson for guidance and his support with the chemicals and equipment used to make our solutions. We would also like to thank Tom Dillon for his guidance and advice throughout the formation of our project. We would like to offer our gratitude to Bryan Winkelman for helping us find references for our project, helping us develop our articles, and creating a website for us to maintain a blog of our research. Finally we would like to extend our immense gratitude to all those who funded and made this project possible: Hong Kim, Patrice Isabella, Tom Bogard, Rob Burkholder, Nicholas Laatsch, Orlando Martinez, Kristen Schurr, and Lori Dishneau.

REFERENCES

1. Bhaya, D., Davison, M., & Barrangou, R. (2011). CRISPR-Cas systems in bacteria and archaea: versatile small RNAs for adaptive defense and regulation. *Annual Review of Genetics, 45*, 273-297.

2. Garneau, J.E., Dupuis, M.E., Villion, M., Romero, D.A., Barrangou, R., Boyaval, P., ... Moineau, S. (2010). The CRISPR/Cas bacterial immune system cleaves bacteriophage and plasmid DNA. *Nature, 468* (7320), 67–71.

3. Jinek, M., Chylinski, K., Fonfara, I., Hauer, M., Doudna, J.A., & Charpentier, E. (2012) A programmable dual-RNA-guided DNA endonuclease in adaptive bacterial immunity. *Science, 337*, 816–821.

4. Ledford, H. (2015). CRISPR, the disruptor. *Nature, 522* (4), 20-24.

5. Leynaud-Kieffer, L. (2016) Relevance of secondary structure for CRISPR spacer. Unpublished master's thesis, University of California Berkeley, Berkeley, California.

6. Mali, P., Yang, L., Esvelt, K., Aach, J., Guell, M., Dicarlo, J., ... Church, G. (2013). RNA-guided human genome engineering via Cas9. *Science, 339* (6121), 823-826.

7. Marakova, K.S., Haft, D.H., Barrangou, R., Brouns, S.J.J., Charpentier, E., Horvath, P., ... Koonin, E.V. (2011). Evolution and classification of the CRISPR–Cas systems. *Nature Reviews Microbiology, 9* (6), 467–477.

8. Reardon, S. (2016) The CRISPR zoo. *Nature, 531,* 160-163.

9. Sinkunas, T., Gasiunas, G., Fremaux, C., Barrangou, R., Horvath, P., & Siksnys, V. (2011). Cas3 is a single-stranded DNA nuclease and ATP-dependent helicase in the CRISPR/Cas immune system. *The Embo Journal, 30* (7), 1335–1342.

10. Sternberg, S.H., & Doudna, J.A. (2015). Expanding the Biologist's Toolkit with CRISPR-Cas9. *Molecular Cell, 58* (4), 568-574.

11. van der Oost, J., Jore, M.M., Westra, E.R., Lundgren, M., & Brouns, S.J. (2009). CRISPR-based adaptive and heritable immunity in prokaryotes. *Trends in Biochemical Science, 34* (8), 401-407.

12. Wang, Y., Cheng, X., Shan, Q., Zhang, Y., Liu, J., Gao, C., & Qiu, J.L. (2014). Simultaneous editing of three homoeoalleles in hexaploid bread wheat confers heritable resistance to powdery mildew. *Nature Biotechnology, 32*, 947-951.

ABOUT THE AUTHORS

Pictured: This was taken near the end of a video conference with Dr. Sarah Richardson. From left to right, there are: Eugene, Dr. Sarah Richardson, and Will.

Over the course of the year, we learned so many things, everything from ordering materials to using CRISPRs to writing journals. Blessed with so much help including all our teachers, mentors, donors, and parents, we were able to take this opportunity to the fullest and actually finish to the end. We're happy to say that the insight we gained in real life research projects reinforced our prospects into the biotechnology field. This research has been a wonderful and an enlightening experience that has helped form us to be better scholars and researchers.

Such an opportunity, especially for high school students like ourselves, is an invaluable asset that will spearhead other opportunities for ourselves in years to come. As we move forward to the universities of our choice, this experience will prove invaluable, especially when we get to do even more lab work in college and beyond. This opportunity has left a deep impression in our hearts, one that enabled us to get a taste of lab work where we could have more free rein, and a lot of fun. We are so glad to have taken this course!

Agrobacterium-mediated transformation of the AtPCS gene into Helianthus annuus (common sunflower)

Naura Taqiya, Cameron B. Reed, Tanya W. Leung, and Shawndra Fordham

Department of Science, Principles of Experimental Design in Biotechnology, Rock Canyon High School, Highlands Ranch, Colorado, USA

The *Arabidopsis thaliana* variant of the phytochelatins synthase gene (AtPCS1), when inserted using *Agrobacterium*-mediated transformation, enables a plant to uptake heavy metals such as cadmium from the surrounding soil. As such, the AtPCS1 gene has been inserted into plant species such as tobacco and Indian mustard in order to determine its viability in phytoremediation; however, the effects of the gene's presence are not well understood. In this research project, we worked towards developing a protocol for inserting the AtPCS1 gene into a *Helianthus annuus* (common sunflower) plant using *Agrobacterium*-mediated transformation. The AtPCS1 gene was inserted into *Agrobacterium tumefaciens* through a heat shock transformation and then introduced into the sunflower plants through both the floral dip method and plant tissue culture. While the bacterial transformation procedure was successful due to its growth in kanamycin-infused agar plates, we cannot confirm the presence of the AtPCS1 gene within the transformed sunflower plants due to time constraints. Such a confirmation can be accomplished by a future group if our research is to continue.

Since the early 1980s, scientists have been working with transgenic plants for a wide variety of different applications such as increased agricultural productivity, phytoremediation, biofuel production, and medicinal purposes[2]. The rapid growth of the plant biotechnology field has also resulted in both a plethora of new methods for inserting genes and an increase in transgenic plant species[2]. One such method for this process is *Agrobacterium*-mediated transformation, which involves the use of the bacterium *Agrobacterium tumefaciens*. *Agrobacterium*-mediated transformation was used to insert the AtPCS gene into the sunflowers. *Agrobacterium tumefaciens* (*Agrobacterium*) is a gram-negative bacterium capable of transferring its DNA into plants, thus making it a popular vector. Since it is a gram-negative bacterium, it has a thinner cell wall than gram-positive bacteria, allowing for easier transfer[7]. *Agrobacterium tumefaciens* can also affect plants ranging from angiosperms to gymnosperms with a high success rate, making it an ideal vector for transfers[13].

Plants transformed using *Agrobacterium*-mediated transformation can occasionally be used for the purposes of phytoremediation, which involves the utilization of plants for environmental cleanup purposes. Phytoremediation is one of the fastest growing areas of research in environmental remediation, and it has resulted in the creation of several different methods for addressing environmental issues[6]. One such phytoremediation method involves planting contaminant-resistant species in areas that have been contaminated by mining, which can help to uptake the contaminants or allow for agriculture to continue in contaminated areas. However, the effects of this process can be affected by several different outside factors such as environment, soil pH levels, and the susceptibility of the contaminants to plant uptake[6].

The gene of interest used in this research project is the *Arabidopsis thaliana* variant of the phytochelatins synthase gene (AtPCS1), which provides plants with protection from heavy metal toxins, most notably from cadmium. The AtPCS gene is naturally found in a wide variety of plants, fungi, nematodes, and algae; however, it is not naturally found in sunflower plants[5]. The PCS gene codes for a type of peptide called phytochelatin, which prevents chemically reactive molecules such as heavy metals from damaging the cell. Due to its properties, the PCS gene has been studied as a potential treatment for soil and groundwater that has been contaminated by heavy metals[6]. The AtPCS1 gene is best activated in the presence of the heavy metal cadmium, but has been capable of remediating heavy metals ions such as arsenate, silver, bismuth, lead, zinc, copper, mercury, and gold[5]. Of the heavy metals, phytochelatins seem to have the best detoxification effect with arsenate and cadmium. Its activation is directly linked to glutathione, an antioxidant important for cell function; low levels lead to low activity while high levels allow for high activity. The AtPCS1 gene has been found to accumulate cadmium through sequestration as molecules bind the metal ions through a process called chelation. After the metal ions undergo chelation, the resulting molecule is transferred into the vacuole, where it combines with other chemicals to form a more complex molecule[4]. Through this process, the heavy metals are integrated into the plant tissue over time, making the plant capable of inducing heavy metal poisoning if eaten in large quantities. However, the absorbed heavy metals have been shown to concentrate in the leaves or roots rather than in the seeds, making the concentration within consumable parts of the sunflower negligible[2]. The AtPCS1 gene has been successfully inserted into tobacco plants using *Agrobacterium*-mediated transformation, making it a plausible gene of interest for insertion into sunflower plants[11]. The *Arabidopsis thaliana* phytochelatin synthase variant (AtPCS1) was used in this research project due to its higher gene expression of cadmium tolerance as compared to the wild-type and other variants[2]. The AtPCS1 gene has been observed to have high sensitivity, reacting with cadmium levels as low as 0.6 micrometers while mutant variants are only reactive at higher concentrations. Previously observed reactions in tobacco (*Nicotiana tabacum*) and Indian mustard (*Arabidopsis thaliana*) include higher survival rates, increased plant growth, and discoloration[2].

The common sunflower plant, named for its tendency to follow the path of the sun, is native to the southwestern parts of the United States but have been cultivated as a staple crop in several parts of the world[9]. Common sunflower plants serve as the most important source of oil within South Africa[12], a nation where over a century of mining has resulted in heavy metal contamination in both water and soil. Examples of heavy metal types found include cobalt, zinc, cadmium, gold, and arsenic[15]. As the common sunflower is both a common crop and susceptible to the crown gall disease[10], it may serve as a good candidate for phytoremediation in historic mining areas such as South Africa.

In this research project, the AtPCS gene (**Fig. 1**) will be inserted into sunflowers using *Agrobacterium*-mediated transformation. The long-term goal is to address mine contamination by developing a common crop with the ability to uptake heavy metals from the surrounding soil and groundwater. This would allow for agriculture to continue in places such as South Africa where industrial and mine contamination had previously hindered it. As contaminated soil has largely contributed to insufficient food production in multiple regions of the world, the development of such a transgenic crop could help to alleviate the issue through phytoremediation.

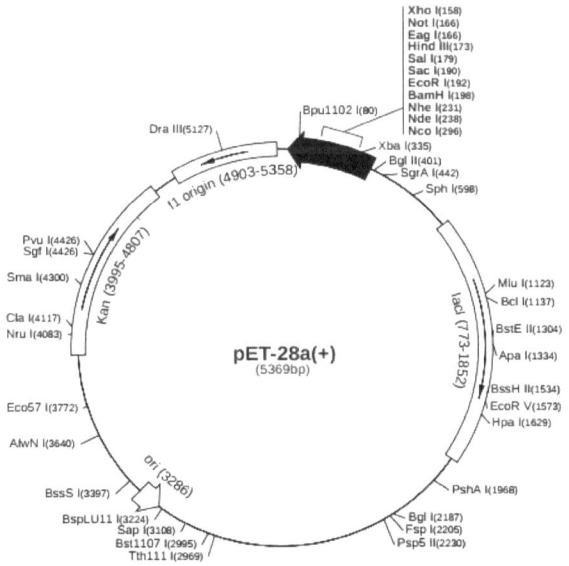

Figure 1. An image of the pET-28a(+) plasmid construct provided by Dr. Joseph Jez of Washington University at St. Louis. The AtPCS1 gene used in our research was first inserted into this plasmid before it was inserted into the pART27 plasmid for use in *Agrobacterium-mediated transformation*. The AtPCS1 placement (signified by the bolded arrow) on the plasmid is close to the kanamycin marker (signified by the Kan label). This plasmid, alongside the pART27 plasmid, both induce kanamycin resistance.

METHODS

The AtPCS1 gene was provided in two types of bacterial vectors: the pET-28a(+) vector (suited for *Escherichia coli*) and the pART27 vector (suited for *Agrobacterium tumefaciens*) (**Fig. 1**). The AtPCS gene contained in the pART27 vector was inserted into *Agrobacterium tumefaciens* cells via heat shock. Two separate methods of plant transformation were then used, as a proper protocol for sunflower has not been established yet.

Preparation for Transformation

Super Optimal Broth with Catabolite repression (SOC) medium was prepared as a growing environment for transformed bacteria. The recipe was used as follows: 95 mL of LB broth, 2.5 mL of 1M KCl, 10 mL of 1 M MgCl$_2$, 10 mL of 1M MgSO$_4$, and 20 mL of 1M glucose[3].

Stock solutions of 100mg/mL amoxicillin and 50mg/mL kanamycin were prepared. MS agar was infused with the kanamycin and amoxicillin stock solutions for plant transformants--the working concentrations of the antibiotics were 50 μg/mL kanamycin and amoxicillin 100 μg/mL when infused with the MS agar (MS+K+Amox).

Vector

The AtPCS1 gene, received from Dr. Joseph Jez with Washington University at St. Louis and Dr. Philip Rea with the University of Pennsylvania, was contained in both a pART27 vector, the *Agrobacterium* counterpart to the pET-28a+ (**Fig. 1**) and a pART27 vector at a concentration of 7.5μg/μl received on filter paper. The former vector is better suited for transformation and propagation in *E. coli* (pET-28a(+))while the latter vector is better suited for transformation in *Agrobacterium tumefaciens* (pART27). The AtPCS plasmid was extracted from the filter paper following a protocol outlined by Dr. Joseph Jez[2]. The plasmid was eluted from the filter paper using 50μL of distilled H$_2$O and centrifuged at 12,000 rpm for one minute. The supernatant contained the plasmid which was further used for transformation. The plasmid was stored at -20°C.

Agrobacterium tumefaciens, received from Carolina Biological, was first cultured to develop pure, rapidly dividing colonies. Approximately 2 μL of *A. tumefaciens* from the received Carolina Biological tube was mixed with 5 mL of LB broth and incubated at 28°C. Cells were then streaked onto sterile LB agar and was incubated at 28°C.

A. tumefaciens Transformation

To transform with the pART27 plasmid containing the AtPCS gene, *A. tumefaciens* cells from the pure colony were made competent with the following procedure. The liquid cultures were centrifuged at 4°C until a pellet was visible. The supernatant removed and the pellet was rinsed with 20mM calcium chloride twice, five minutes each at 4°C. The resulting *A. tumefaciens* mixture was then stored at -20°C.

A. tumefaciens was transformed using standard heat shock protocol from AddGene[1] using a 28°C incubation temperature and a 42°C heat shock temperature. 10 μL of supernatant containing competent *A. tumefaciens* cells and 50 μL of the AtPCS solution were mixed together and stored on an ice bath. The mixture was moved into a hot water bath heated at 42°C for 45 seconds, then immediately placed back into the ice bath. 50 μL of SOC medium was added and the resulting mixture was placed on the shaking incubator for 2 hours at 28°C at 200 rpm. After two hours had passed, the transformed *A. tumefaciens* were spread onto kanamycin infused petri plate and returned to the 28°C incubator without shaking. Varying volumes of transformants were plated onto different kanamycin plates (5μL, 10μL, 25μL, 50μL, and 450μL) to test transformation efficiency. Non-transformed *A. tumefaciens* bacterium were also streaked onto kanamycin infused agar to serve as a control.

Plant Transformation

Sunflower seeds (*Helianthus annuus*) obtained from Jonsteen Company were planted and used in this experiment for transformation **(Pic. 1a)**. Watering holes were inserted the the bottom sides of the cup before filled with soil and planted. A total of 35 sunflower seeds were planted, approximately one inch deep **(Pic. 1b)**. Separate sunflowers seeds were also germinated manually. Five seeds were placed in 50% bleach for five minutes and after the five minute period, the seeds were rinsed in sterile water containers for five minutes, four consecutive times. The seeds were then vertically submerged into a vial of multiplication media (originally designed for African Violets from Carolina Biological, but has proven to grow sunflower tissue).

Picture 1: These are two pictures of the sunflower plants before the transformation processes. **(a)** A picture of the sunflowers being fertilized from their position at the light cart. **(b)** A picture of thirty of the thirty-five cups used for planting the sunflower seeds. The sunflower seeds have just been planted in the soil.

The transformed *A. tumefaciens* demonstrated healthy growth after approximately three days of incubation. We then began transformation of *Helianthus annuus* with the AtPCS gene. Two methods of transformation were used, floral dip and explant *A. tumefaciens*-mediated transformation.

In our floral dip protocol, the floral dip mixture was comprised of 125 μL of 0.0025% Silwet-L77 and 200 mL of transformed *A. tumefaciens* solution. The upper half of a flowering *H. annuus* was dipped into a Silwet L-77, LB broth, and *A. tumefaciens* mixture then enclosed in complete darkness for approximately 12 hours **(Pic. 2)**. The sunflowers were then reintroduced back to light cart to growth. As a precaution, weed barrier was used to cover the soil.

Picture 2: In the first plant transformation trial, sunflowers were transformed using the floral dip method. After the dip, the sunflowers were placed in a container to prevent contamination and transferred into complete darkness for the next 12 hours.

In our second method of transformation, explants were derived from varying forms of the sunflower, including a germinated seed, the apical meristem, and leaves selected from the mid-section of the sunflower plant[⁴]. They were sterilized by soaking in soapy water for 4 minute, rinsing in DI water for 2 minutes, and soaking in 70% isopropyl alcohol for 2 minutes, followed by a rinse in sterile water for 5 minutes. Lastly, they were soaked in 10% bleach for 8 minutes and then rinsed twice in sterile water for 5 minutes. The sections of the a sunflower leaf were cut into explants. Five were placed in the *A. tumefaciens* floral dip mixture for transformation and two leaf explants remained untransformed to serve as a control. A total of seven explants from the sunflower leaf were placed in two separate petri dishes, one for control and one for experimental, containing MS+K+Amox medium **(Pic. 3b)**.

Picture 3: These pictures demonstrate the conditions of the sunflowers and explants after the plant tissue culture or floral dip processes. The first picture is of one of the two sunflowers transformed using the floral dip method **(a)**. The forming seeds are barely visible in the center of the sunflower. The next picture is of the five explants transformed using plant tissue culture **(b)**. The explants are visibly discolored and there is heavy condensation within the agar plate. The last picture is of the apical meristem and explants transformed using plant tissue culture **(c)**. All four tubes show no signs of contamination but do not have any visible growth at the time.

Four other explants derived from leaves and one apical meristem were all similarly transformed and then placed vertically in the MS+K+Amox medium, but in tubes instead of petri dishes. In order to transform the germinated seed, a small vertical incision was made before dipping it into the *A. tumefaciens* transformant and similarly placed is MS+K+Amox tubes. Two explants from sunflower leaves were cut and untransformed to serve as a control for the tubed variants (**Pic. 3c**).

RESULTS

Transformation success of the *A. tumefaciens* bacterium were measured by their growth and was not verified by using PCR. Control trials wherein a pure, untransformed *A. tumefaciens* colonies were placed on kanamycin infused LB plates no growth was detected; however, the transformed *A. tumefaciens* experienced successful growth in the presence of kanamycin. The transformation efficacy was calculated using the following formula:

$$Transformation\ Efficacy\ \left(\frac{transformants}{\mu g}\right) = \left(\frac{\#\ colonies\ on\ plate}{ng\ of\ DNA\ placed}\right) 1000\ \frac{ng}{\mu g}$$

The transformation efficacy was relatively consistent regardless of the amount of transformants placed on the plate, with the notable exception of the volume of 10 μL (**Table 1**).

Transformation Efficiency of AtPCS Gene in *A. tumefaciens*		
Volume of Transformants (μL)	Number of Colonies	Transformation Efficiency (colonies / μg)
5	3	0.176
10	18	0.528
25	14	0.164
50	25	0.147
450	272	0.177

Table 1: This table shows the transformation efficiency compared to the volume of transformants spread on the plate.

Both of the experimental sunflowers demonstrated continued growth and began to flower after the floral dip (**Pic. 2**). A bud began to wither on one of the flowers, but another bud grew beside it. Since the flower is beginning to bloom, the sunflowers are expected to set seeds soon.

Both experimental and control explants in the petri plate have shown discoloration (**Pic. 3**). Explants in the tubes have shown no signs of contamination, but have not shown any signs of growth either.

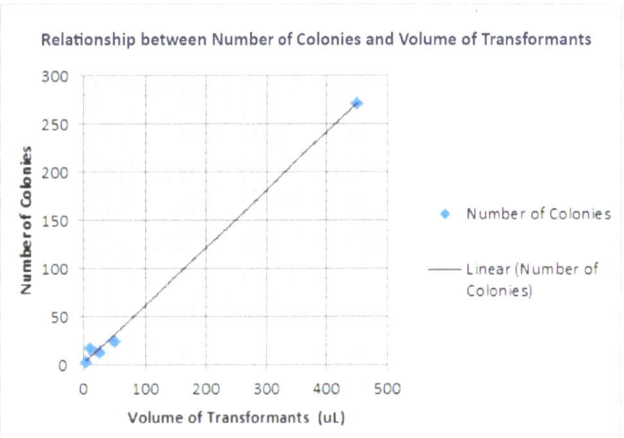

Graph 1: This graph depicts the volume of transformants (uL) spread on the plate versus the number of *Agrobacterium tumefaciens* colonies cultivated on kanamycin-induced agar plates.

DISCUSSION

The overall objective of our research was to begin the process of developing a crop capable of uptaking heavy metals from contaminated soil and groundwater, allowing for its use in the phytoremediation of areas contaminated by mining practices. Due to time constraints, we focused on developing a protocol for successfully inserting the AtPCS1 gene into *Agrobacterium tumefaciens*. In order to achieve this, our team conducted three separate attempts at inserting the plasmid using heat shock protocols based on previous research articles. Based on the level of success we observed in the first two trials, we worked with our mentors and adjusted our protocols in order to find a viable procedure for *Agrobacterium tumefaciens*. Typically, electroporation (running a current through the bacteria to make it competent) would be preferable for use in bacterial transformation due to its higher efficiency rate of up to 10^{10} transformants/ug[14]; however, we were unable to use electroporation due to a lack of materials and equipment. Heat shock transformation with the $CaCl_2$ method, on the other hand, is a relatively simple process requiring minimal materials and less time to execute, making it one of the more convenient methods[14].

We have evidence to suggest that we successfully transformed the *Agrobacterium tumefaciens* with the pART27 plasmid containing the AtPCS1 gene on our third attempt, as the bacteria experienced growth on kanamycin-infused agar plates. Based on the data we collected, there appears to be a strong positive correlation (r= 0.998) between the volume of transformant (in uL) and the number of *Agrobacterium tumefaciens* colonies growing on the agar plates, and the percent variability of the data is r^2=0.997, indicating that the variability is focused around the mean-- the regression line fits the data (**Graph 1**). In context, the data represented in the table and the graph indicate that the growth of the *Agrobacterium* in the kanamycin-induced agar plates follows a roughly consistent pattern of growth, yielding transformation efficiencies that are fairly close to each other (**Table 1**). However, it can be noted that an outlier exists with the 10 uL transformant value and that an influential point exists with the 450 uL transformant value;

such data points can cause the correlation and variability to appear stronger than they actually are. In this case, the correlation and percent variability remain moderately positive with the two points omitted, suggesting that the bacterial transformation procedure was successful and consistent overall. In addition, the transformed *Agrobacterium tumefaciens* could be successfully isolated in a mixture of kanamycin and MS broth, as the resulting mixture demonstrated significant turbidity rather than clarity.

In our attempt to insert the AtPCS1 gene into sunflower plants, we utilized two different methods of *Agrobacterium*-mediated transformation: the plant tissue culture method and the floral dip method. We chose to attempt two different methods because while the sunflower plant had been transformed before, it had never been transformed with the AtPCS1 gene in the past; as a result, we attempted to standardize a protocol for the sunflower plant. Plant tissue culture is typically used for the process of transformation with the sunflower plant[s], but there was not much information on what parts of the sunflower were ideal for transformation; in other plant tissue culture protocols, a leaf disc was the best part to use with an apical meristem as a viable alternative[8]. Therefore, we performed plant tissue culture using an apical meristem, leaf discs, and germinated sunflower seeds (**Pic. 2c**). There is little to no research on the floral dip method's success with the *Agrobacterium*-mediated transformation of sunflowers. The floral dip has, however, been successfully performed in *Arabidopsis thaliana* plants, and as a result, we performed a floral dip on two of our budding sunflowers for the purposes of comparison with the plant tissue culture method (**Pic. 2a**).

As of recent, there has been no contamination present with the plant tissue culture sunflower parts, but there is also no visible growth (**Pic. 2b**). However, some of the explants do appear to possess a brown discoloration, indicating the possibility of excess sterilization and resulting cell death. As the plant tissue culture process involves the use of bleach to eliminate possible contaminants on the explants, it is possible that the explants may have been soaked in the bleach for longer than required. The sunflower seeds, on the other hand, may have experienced a lack of growth and contamination due to their inviability, as five of the thirty-five sunflower seeds planted in the soil did not grow. These results indicates that sterilization times may need to be adjusted to produce a viable protocol.

Due to time constraints, we have been unable to confirm the presence of the AtPCS1 gene inside both the floral dip sunflowers and the sunflower explants. Therefore, the next step in our research would be to run a polymerase chain reaction (PCR) on the transformed sunflower plants. In order to run a PCR, however, two different procedures can be performed with the floral dip sunflowers and the sunflower explants. The seeds produced by the floral dip sunflowers cannot be used to run a PCR in their current form--the seeds would need to be grown into another set of sunflower plants first. The sunflower explants, on the other hand, would need to develop into calluses before a PCR can

be run on them. However, it would be preferable for the sunflower explants to undergo multiple generations in order to accurately confirm the AtPCS1 gene's presence.

ACKNOWLEDGMENTS

We are grateful to have received help and support from a wide variety of individuals throughout our research, and we would like to take the time to thank everyone who helped make our project happen. We would first like to give a special thanks to Uma Venkitanarayanan, a Rock Canyon High School parent and former researcher for all her help and support with developing and executing experimental protocols and supervising our research. We would also like to give a special thanks to Joseph Jez from Washington University at St. Louis and Philip Rea from the University of Pennsylvania for providing us with the AtPCS1 gene and the pART27 plasmid that we used in our research, and to Fred Moshiri and Sunny Gilbert of Monsanto and Kevin Lutke of the Danforth Plant Science Center for helping us design our research project. In addition, we would like to thank Diane Keely, Jeff Dellin, and Nelson Freeman for making the donations that enabled us to conduct our research. We would also like to thank Amy Hacker and Susanne Petri for providing classroom space, Rock Canyon High School for providing laboratory space and equipment, and the Douglas County School District and Sherri Bryant for providing funding for the advanced equipment used in our experiment. Last but not least, we would also like to thank Tom Dillon from the Community College of Aurora for supporting and providing initial feedback on our project, David Ferguson of Rock Canyon High School for providing many of the chemicals and equipment, and Bryan Winkelman for providing guidance throughout the entire experiment and for helping with the publication of our research.

REFERENCES

1. Bacterial Transformation. (2013). *Addgene*. Retrieved 2015, November 11. [Web]
2. Cahoon, R., Lutke, W., Cameron, J., Chen, S., Lee, S., Rivard, R., . . . Jez, J. (2015, May 27). Adaptive Engineering of Phytochelatin-based Heavy Metal Tolerance. *Journal of Biological Chemistry J. Biol. Chem.*, 17321-17330.
3. Chan, W.-T., Verma, C. S., Lane, D. P., & Gan, S. K-E. (2013). A comparison and optimization of methods and factors affecting the transformation of *Escherichia coli*. *Bioscience Reports*, 33(6)
4. Chabaud, M., Ratet, P., De Sousa Araújo, S., Lopes, A., Duque, A., Harrison, M., & Barke,r D. (2007). *Agrobacterium tumefaciens*-mediated transformation and in vitro plant regeneration of *M. truncatula*. *Medicago Truncatula Handbook*, 1-34.
5. Cobbett, C. (2000). Phytochelatins and Their Roles in Heavy Metal Detoxification. *Plant Physiology, 123*(3), 825-832.
6. Dzantor, E. & Beauchamp, R. (2002). Phytoremediation, Part I: Fundamental Basis for the Use of Plants in Remediation of Organic and Metal Contamination. *Environmental Practice*, 1(2), 77-87.
7. Gelvin, S. (2003, March 3). *Agrobacterium*-Mediated Plant Transformation: The Biology behind the "Gene-Jockeying" Tool. *Microbiol Mol Biol Rev. Microbiology and Molecular Biology Review*, 67(1), 16-37.
8. Mohmand, A., & Quraishi, A. (1994). Tissue Culture of Sunflower. *Pakistan Journal of Agricultural Research*, 15(1).
9. National Sunflower Association : Frequently Asked Questions. (2015). *National Sunflower Association*. Retrieved 2015. November 9. [Web]
10. Nikneshan, P., Karimmojeni, H., & Moghanibashi, M. (2011). Allelopathic potential of sunflower on weed management in safflower and wheat.*Australian Journal of Crop Science, 5*(11), 1434-1440.
11. Romanyuk, N.D, Rigden, D.J, Vatamaniuk, O.K, Lang, A, Cahoon, R.E, Jez, J.M, Rea, P.A (2006) Mutagenic definition of papain-like catalytic triad and sufficiency of N-terminal domain for single-site ica – where climate change may trigger a toxic time bomb. *The Guardian*. Retrieved 2015, November 9. [Web]

ABOUT THE AUTHORS

Pictured: Our mentor Uma Venkitanarayanan, Naura Taqiya, Cameron Reed, and Tanya Leung (left to right). Mentors-- Dr. Joseph Jez with Washington University at St. Louis, Dr. Fred Moshiri, Dr. Sunny Gilbert with Monsanto, and Dr. Kevin Lutke with the Danforth Plant Science Center are not pictured.

Over the course of the last ten months, we have learned many things about developing and executing scientific research. The most valuable lesson we have learned is resilience--our project started as a mere shred of hope. We were often told that there was no way it could be done: in the starting months our project changed more times then we entered the classroom. However, thanks to Uma Venkitanarayanan we were able to pull off this great attempt. Under her guidance, we were able to acquire many invaluable lab skills; from something as basic as preparation of basic stock solutions to great feats like genetic alterations, there. We also had the continued support of Dr. Joseph Jez--who provided us with the AtPCS1 gene, bacterial expression vector, and the background information we needed in order to use it-- Dr. Fred Moshiri, Dr. Kevin Lutke, and Dr. Sunny Gilbert. They all helped us to prepare the research proposal and background information needed to get our research started.

Cumulatively, it has shown us how a real-life lab works, demonstrating both the ups and downs of research. The class overall has been the best classroom experience we have ever had. We have learned more in a class where the instructor doesn't teach us, but rather we learn all the materials ourselves. It's those struggles that make the knowledge that much more enduring.

This course has inspired us to continue pursuing scientific subjects: All of us plan to continue our studies in the sciences as we move onto college. Cameron will be studying Plant Biology with a concentration in Molecular and Biochemical Physiology; Tanya will be focusing on either pre-medicine or environmental engineering; and Naura will be majoring in in Genetics and Molecular Biology.

The effect of freezing on bacterial growth found on Levi's® denim

Kathryn C Smith, Seana P Thompson, Benjamin J Huxley, and Shawndra L Fordham

Department of Science, Principles of Experimental Design in Biotechnology, Rock Canyon High School, Highlands Ranch, Colorado, USA

This study investigated the effects of freezing on Levi's® denim in order to evaluate the claim that freezing the jeans for 24 hours will kill the microbial growth. Patches of denim were worn for 28 days after which half were frozen at -20°C while the other half were left untreated. Next the bacteria was cultured and the DNA was extracted and sequenced. Bacterial populations were identified, their relative abundance was calculated, and a test of significance was performed comparing the two treatment groups. Our results show that there is no statistical difference between the amount of bacteria present on the jeans between the frozen and unfrozen treatment groups. This suggests that freezing the jeans does not have a significant impact on microbial growth and Levi's claims are inaccurate.

Levi's jeans' CEO, Charles Bergh, has claimed that people do not need to wash their jeans to remove microbial growth and that freezing the jeans is just as effective as putting them in the wash. We believe that freezing Levi's denim will not eliminate all microbial growth because extremophiles that thrive in cold climates may go dormant and then begin to repopulate the jeans after they thaw. We also believe that because freezing the denim does not actually dislodge the skin cells or other possible particles that bacteria may grow on, simply freezing the jeans could result in rapid microbial growth post-freezing due to the abundance of microbial food sources. Extremophiles are able to exist and thrive in environments that include very high or low temperatures, in concert with toxic chemicals, radiation, and areas with high pressure. The lowest temperature recorded for active microbial communities is -18°C.[12] Organisms that are known to thrive at 15°C or lower are known as psychrophiles. Specific types of psychrophiles include *Pseudomonas*, *Listerias*, *Vibrios*, and *Coliforms*. Some of these bacteria are harmful to humans and they are known to result in meningitis, sepsis, food poisoning, and diarrhea.[10]

If freezing jeans proves to stop microbial growth, then we as a society could save water by not washing jeans or other clothing and simply freezing them. Jeans are one of the few pieces of clothing worn by a large number of the people on this planet and a breakthrough in this could have a major impact. It can change how people wash, wear, and think about their jeans. This project will also be able to help contribute to the rapidly expanding human microbiome project by adding new species that may have not been there before.

Microorganisms play many roles in all aspects of life. They are a known cause of disease but also play a role in atmospheric chemistry, digestion in large organisms, the breakdown of organic matter, and global geochemical cycles.[3] Microbe communities interact in their environment in many ways and influence - and are influenced by - their surroundings. Beyond this, very little is known about microbes. One difficulty of studying microbes is that their complete role/function in an environment can only be seen when they are in the environment interacting with other microbes. Due to this, when they are taken into a lab and isolated, it becomes difficult to identify the species and its environmental role.[1]

Some of the most common bacteria found on humans and therefore some of the bacteria we expect to find in our research are in the genera *Corynebacterium*, *Staphylococcus*, *Micrococcus*, and *Bacillus*, *Corynebacterium* are known to grow in the moist folds parts of the human body, such as armpits and groins.[6,9] *Staphylococcus* is commonly found on the skin and in mucous areas in the nose and is typically harmless. *Micrococcus* is responsible for the smell associated with sweat. *Bacillus* has been found causing abscesses, urinary tract infections, and respiratory infections.[14] Growing certain bacteria can be difficult, as it is practically impossible to mimic the conditions of the skin.[8] However, it is likely that if bacteria survive the freezing process the most common bacteria found on the skin of humans will be found.

Levi's originally claimed said that instead of washing your jeans, jeans should be frozen and this would kill the bacteria. If this claim proves to be false, people could be wearing pants with continuously growing colonies of bacteria. If this is the case this would not be sanitary and could have the potential to be a health risk. Levi's jeans has removed this claim from their website since Dr. Craig Cary's interview with the Smithsonian regarding his doubts over the effectiveness of freezing jeans.[15] However, the fashion website Elle.com confirmed that Levi's quickly returned their phone call confirming that they believe freezing your jeans would kill the bacteria. On the website, Levi's VP of Women's Design, Jill Guenza, stated that freezing jeans is an easy and efficient alternative to washing the jeans, which can cause them to fade and shrink.[5] There are no other jean companies that have claimed that freezing jeans makes them sterile. However, there are many suggestions on the Internet that advise washing your jeans at most twice a year. Also, Tommy Hilfiger commented that it is crazy to wash jeans after every wear.[11] Levi's claim is not backed by scientific evidence. False advertising is an issue in the United States and companies' claims should be investigated for accuracy. One example of false advertising occurred when Splenda claimed their products were made with real sugar, when in reality it was an artificial chemical.[13] Also, Eclipse gum claimed that their gum could help kill germs in your mouth when in reality, it could not.[2]

In our experiment, we will be wearing denim patches against our skin for 28 days and then freezing half of each patch while leaving the other half at room temperature in order to compare the differences in the amount of microbial growth present. Bacteria on each treatment patch will be cultured and bacterial DNA will then be extracted, amplified using PCR, and then sent to the Waikato Lab in New Zealand for sequencing.

METHODS
Our team purchased a new pair of 501 Levi's Original jeans and cut all of the denim patches for our entire experiment from this pair of jeans.

Experimental Design
In our preliminary trial, two females and one male wore one 16x2 in² denim patch pinned around each individual's upper thigh for two weeks (8 to 12 hours daily). They were not worn when sleeping or bathing. Each individual kept a

16x12 in² patch in their closet at room temperature as a control. After wearing these patches for two weeks, bacteria was removed from the denim using 10 mL 1X PBS and 250 μL 250 μL of PBS/bacteria solution from each patch was cultured on MP Biomedicals LB Agar petri plates. The plates were cultured at 37°C until optimal growth was observed (**Pic. 1**). From this trial, it was decided that the best denim patch size to use to obtain optimal bacterial growth would be a 3x2 cm²

Picture 1: Kathryn and Seana preparing denim patches in the Biological Safety Cabinet

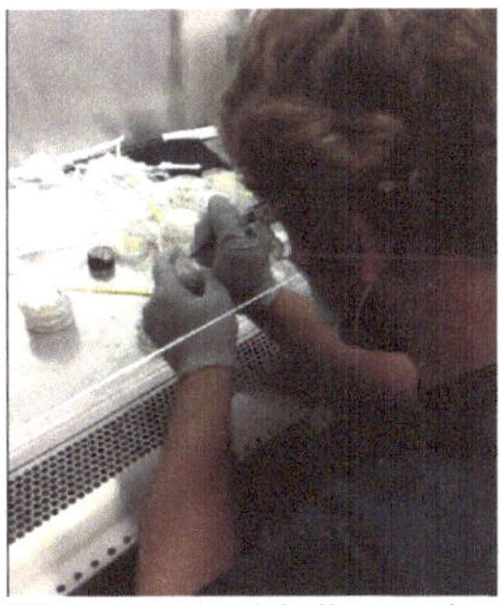

patch with 250 μL of PBS/bacteria solution added to the petri plate for optimal bacterial growth (**Pic. 2**).

Next, we began our experiment. Two males and two females each wore three 3x2 cm² denim patches pinned to the inside of their clothing on their upper thighs everyday - all day, except when sleeping or bathing for 28 consecutive days. During this time, all four subjects also had a control denim patch in their closets exposed to the air and a control denim patch

Picture 2: Ben labeling petri plates to grow the bacteria from our patches.

attached to the outside of their backpacks.

Bacterial Culturing
After the 28 day period, the bacteria on the denim patches were cultured. The bacteria were cultured prior to DNA extraction to ensure the extraction of DNA from living bacteria and to avoid the accidental extraction of residual DNA. First, one half of each denim patch was frozen at -20°C for 24 hours. After 24 hours, the bacteria were washed from all denim patches so that they could be cultured. To wash the bacteria from the denim patches, the denim patches were placed in 10 mL of sterile PBS solution of a 1X concentration and left to soak for 30 minutes on a rocking platform. To culture the bacteria, 250 μL of PBS/bacterial solution was spread onto MP Biomedicals LB Agar petri plates. Finally, the plates were incubated at 37°C until optimal growth was observed. There was optimal growth on almost all of the petri plates.

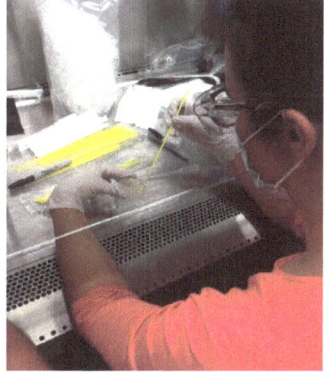

Picture 3: Petri plates, the top plate showing optimal growth, the middle plate being slightly overgrown, and the bottom plate with no growth.

Optimal growth was determined when colonies were visible yet had not grown enough to be touching each other (**Pic. 3**). After 24 hours, all petri plates were removed from the incubator, regardless of growth. The plates were stored at 4°C to stop microbial growth.

After all plates were removed from the incubator, the bacteria were washed from the petri dishes using 0.5 mL of sterile water and an L spreader to loosen the colonies before removing the sterile water and placing it into a microcentrifuge tube. This was done three times for each plate and all three 0.5 mL aliquots from the same plate were placed into the same microcentrifuge tube. After washing the bacteria, these tubes were stored at -20°C (**Pic. 4**).

Picture 4: Using an inoculation loop to loosen bacterial colonies off a bacteria culture.

DNA Extraction and Amplification
The next step was to extract the DNA from the cultured bacteria. DNA extraction protocols are outlined in the MoBio Biostatic Bacteremia DNA Isolation Kit (12240-50). After extracting the DNA, PCR was performed on each sample. Primers obtained from Dr. Craig Cary's lab at University of Waikato were resuspended using 2.5 μL of

ultra pure water and the rehydrated primers, were stored in the -20°C freezer.

Two PCR reactions were performed on each sample using PCR Ready Beads, 2 μL of DNA, 0.5 μL of the forward primer, 0.5 μL of the reverse primer, and 23 μL of sterile H20. The thermocycler program used was 94°C for 3 min of denaturation and then 30 cycles of 94°C for 45 s of denaturation, 50°C for 1 min of annealing, and 72°C for 1.5 min of extension phase. After 30 repetitions, it was at 72°C for 10 min of extension phase and then was held at 4°C.

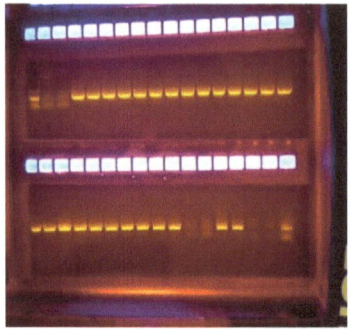

To verify successful amplification, gel electrophoresis was performed using a Lonza 1.2% fast gel (**Pic. 5**). After checking the success of the PCR reactions, the two reactions for each sample were combined back into one sample tube.

Picture 5: A gel confirming the success of PCR. Ladder in well 1, well 2-32 hold samples 47-79 in order.

Sequencing and Data Analysis
The samples were dried using a heating block in a Biological Safety Cabinet. The dried samples were then shipped to Dr. Craig Cary's lab at the University of Waikato, New Zealand, where each sample was sequenced using Ion Torrent Technology. Data were analyzed using tests of significance and by creating tables to interpret and show our results.

RESULTS
DNA sequencing enabled us to identify species and relative abundances of bacteria present for each treatment group, frozen and unfrozen. Then, t-tests were used to determine the type of bacteria most prevalent in each treatment and reveal a statistical difference in the bacteria present and between the two treatment groups.

The relative abundance for each genus of bacteria found on the jean patch was calculated. To determine relative abundance, the number of reads from a specific genus was divided by the total number of reads for that sample. We then created a table showing the relative abundance of each genus in the different treatment groups. The table also shows whether the patch was kept in a closet, kept on a backpack, or worn on a leg (**Table 1**). In all groups, there were a large amount of genera with low relative abundance and a few genera with high relative abundance. The patches worn on the legs (both frozen and unfrozen) had a much higher number of reads than the patches kept in closets or on backpacks (**Graph 1**). This shows that the amount of bacteria on the denim patches worn against legs had more bacteria than the others. However, the number of reads between frozen and not frozen denim patches, were similar, showing they had similar amounts of bacteria present.

Relative Abundance of Genera Found on Denim						
*denotes genus known to be a human pathogen						
Genus	Backpack Frozen	Backpack Unfrozen	Closet Frozen	Closet Unfrozen	Leg Frozen	Leg Unfrozen
Acidovorax *	0.092%	-	-	-	0.001%	-
Acinetobacter *	6.399%	0.448%	36.626%	3.716%	0.717%	0.518%
Aquabacterium	-	-	-	0.008%	0.030%	0.023%
Aquimonas *	0.061%	0.050%	0.069%	0.086%	0.088%	0.095%
Bacillus *	3.699%	3.545%	3.695%	5.257%	5.643%	5.060%
Burkholderia *	0.079%	0.079%	0.108%	0.130%	0.150%	0.123%
Cloacibacterium	0.007%	11.185%	-	-	-	-
Corynebacterineae *	0.120%	0.868%	0.437%	3.600%	0.284%	0.826%
Deinococcus	0.039%	-	0.051%	0.157%	0.023%	0.080%
Ferruginibacter	0.083%	-	0.007%	0.033%	0.004%	-
Fulvimonas	0.054%	-	0.020%	0.006%	0.032%	0.091%
Gemella *	1.588%	0.308%	0.005%	0.028%	0.107%	0.002%
Haemophilus *	-	-	-	0.446%	-	-
Lactobacillus *	4.783%	20.244%	1.506%	0.960%	0.253%	0.616%
Methylobacterium *	0.008%	-	-	4.341%	-	0.034%
Micrococcineae *	20.770%	9.333%	0.676%	0.292%	0.636%	0.647%
Paracoccus	-	-	-	0.093%	-	-
Prevotella *	0.245%	0.068%	0.011%	0.054%	0.007%	0.003%
Pseudomonas *	0.083%	-	0.006%	0.012%	0.005%	0.047%
Pullulanibacillus	0.014%	0.012%	-	0.423%	4.582%	3.730%
Sphingobium	-	0.008%	-	0.016%	0.004%	5.982%
Sphingomonas *	0.021%	-	0.016%	1.040%	0.015%	0.299%
Staphylococcus *	60.938%	53.692%	56.512%	78.901%	87.136%	81.571%
Streptococcus *	0.057%	0.036%	0.053%	0.061%	0.081%	0.068%

Table 1: This table shows the relative abundance of bacterial genera found on the denim patches of each treatment. Only those genera that had higher abundances were shown in this table. A dash indicates a value of zero. Genera denoted with a * are associated with human disease.

Graph 1: This graph shows the number of bacterial sequences (reads) that were obtained through sequencing for each treatment.

We performed t-tests comparing frozen versus unfrozen denim patches. The t-tests gave us a p-value of .9565 which showed the variability in bacterial abundance between frozen and not frozen denim patches was not statistically significant.

DISCUSSION
The primary purpose of this research was to determine if freezing Levi's denim resulted in the elimination of microorganism present on the denim. Our t-tests all failed to reject our null hypothesis - that there is no difference between the amount of bacteria present on unfrozen versus frozen denim patches. This means that, from a statistical standpoint, we did not find sufficient evidence to confirm that freezing results in a difference in microbial growth on denim. From the standpoint of our experiment, we found evidence supporting our hypothesis that freezing jeans does not actually remove microbes from jeans.

As the t-tests we performed gave p-values that were above .05, we have evidence that there is no significant difference between the treatments of freezing and the control, thus we failed to reject our null hypothesis. The denim patches worn against the leg were found to have the highest quantity of bacteria present while the patches kept on the outside of the backpack throughout the experiment were found to have the lowest quantity of bacteria. The genera of bacteria that were most prevalent were *Staphylococcus*, *Bacillus*, *Acinetobacter*, *Sporosarcina*, and *Corynebacterineae*. *Staphylococcus*, *Bacillus*, and *Corynebacterineae* were expected as they are commonly found on skin.[8] *Acinetobacter* and *Sporosarcina* were not anticipated to have such high abundance on the patches because they are most commonly found in soil and water.[1] A greater diversity of bacteria was found on frozen denim patches than on unfrozen denim patches. We suspect this is due to the freezing treatment weakening the usually dominant genera of bacteria and allowing less dominant genera the opportunity for more growth than would be possible under normal conditions. Furthermore, the frozen backpack denim patches fostered the greatest diversity of bacterial genera and this may be due to their exposure to the greatest number of different environments. After freezing, these diverse bacteria were given the opportunity to thrive as they did not face competition from the usual dominant genera.

Some of the most prevalent genera found on the denim patches have species that are known to cause diseases in humans. All species of *Acinetobacter* have been found to cause illness according to the CDC, including cases of pneumonia or blood infections.[1] *Staphylococcus aureus* is well known to cause many dangerous skin infections, but other species of the genus can cause urinary tract infections and toxic shock syndrome.[6] Despite the fact that many of the genera found on our denim have pathogenic species, the majority of strains of these genera do not harm humans.

Based on our research, there is clear evidence that refutes Levi's claim that freezing your jeans for 24 hours will "clean" them by killing microbes present. This opens the door for other researchers interested in further testing this claim. Besides using science to test advertising claims, our research can contribute to the science world's knowledge of individual jean patches. It is important to note that not every species of bacteria was represented through the culturing, extraction of DNA, and the rest of our data collection. The majority of the organisms on the human skin do not grow well in cultures.[3] Only some tolerant, resilient species are able to grow on a medium other than human skin. For example, *Staphylococcus* is known to grow on both skin and in culture as well. Furthermore, the denim patches worn by the individuals only made contact with a small portion of their skin and therefore would not have collected the entire microbiome. Despite the fact that many bacteria found on skin may not have grown, this experiment gave a basic overview of the effects of freezing on microbial growth on denim and bacteria present in our everyday environment, as well as bacteria present in more extreme environments.

There were many differences between the microbes found on each individual. The differences in dietary habits, genetics, living condition, and personal hygiene routines from person to person may have led to the differences in bacteria species and relative abundance found on each denim patch.

Our research is not fully comprehensive because of several errors that occurred throughout our experiment. One major error that occurred was the accidental washing of a sample set of denim patches one week before the experiment ended. To make up for washing the jeans, the patches were worn at night as well as during the day for the remaining week of the experiment. This resulted in less bacteria than there would have been otherwise and lessened the amount of data collected. As we are looking at the difference between living bacteria on frozen versus unfrozen patches from the same original denim patch, our conclusions are unchanged despite this error. We also had one issue with labeling samples that caused a mix-up of three frozen samples and we were unable correctly identify which was which with certainty. This means our graph comparing the number of reads between the location and treatment could be inaccurate. If looking to compare results from person to person, this error eliminates these three samples from the pool of data as we do not know which sample is which.

To continue this research, the next step would be to compare the differences between washing and freezing denim, as compared to no treatment. This could give a basis for what microbe-free denim is and allow us to see if washing has a larger impact on growth as compared to freezing. To truly refute the claim made by Levi's, it would be necessary to know how washing affects the microbial growth and how that growth compares to the growth on frozen denim. Another test would be to continue wearing the denim for a second month after treatment to see if there is an increase in microbial growth after months of not washing it.

ACKNOWLEDGMENTS
We would like to thank Dr. Craig Cary, Professor of Biological Sciences and Director of DNA Sequencing at University of Waikato, for mentoring us with our research, guiding the research question, and providing us with materials, as well as his lab technician, Roanna Richards-Babbage, for answering our questions and supporting us throughout our project. We would like to say a special thank you to the following people for funding this research: Teresa Hull, Amy Antinoro, Mark Grafitti, Lori Dishneau, Kristen Schurr, Tom Bogard, Rob Burkholder, and Nicholas Laatsch. Also, thank you to Amy Hacker for assisting with experimental design questions and other lab questions. Thank you to Ruth Smith for aiding us with data analysis and the creation of graphs. We also need to thank Tom Dillon for his feedback and support with the design and implementation of our project. Another thank you to David Ferguson for providing help with chemicals and other laboratory needs. We also need to thank Susanne Petri for sharing her laboratory space with us, Rock Canyon High School for providing laboratory space and equipment, Douglas County School District for the Innovation, and Perkins Grant funding that provided research grade laboratory equipment necessary for this experiment.

REFERENCES

1. Acinetobacter in Healthcare Settings. (2010, November 24). Centers for Disease Control and Prevention. Retrieved 2016, April 18. [Web]

2. Bhasin, K. W. (2011). 14 False Advertising Scandals That Cost Brands Millions. Retrieved 2016, April 19. [Web]

3. Coil, D., & Eisen, J. (2011, June 3). Fact Sheet: Microbial Ecology in the Built Environment. Microbiology of the Built Environment network. Retrieved 2015, October 8. [Web]

4. Grice, E. A., & Segre, J. A. (2013, January 3). The skin microbiome. National Center for Biotechnology Information. Retrieved 2016, March 31. [Web].

5. Hoff, V., & Levinson, L. (2013, November 13). Why You Should Store Your Jeans in the Freezer. *Elle*. Retrieved 2015, October 8. [Web]

6. Kloos, W., & Musselwhite, M. (1975). Distribution and Persistence of Staphylococcus and Micrococcus Species and Other Aerobic Bacteria on Human Skin. *Applied Microbiology, 30*(3). National Center for Biotechnology Information. Retrieved 2015, September 29. [Web]

7. Life In Extreme Environments. (2015). University of Utah Health Sciences. Retrieved 2015, October 3. [Web]

8. MacLachlan, A. (2012). Body Bacteria Science Publication. National Institute of General Medical Sciences. Retrieved 2015, October 8. [Web]

9. Micrococcus. (2011, April 19). Public Health Agency of Canada. Retrieved 2015, October 3. [Web]

10. Microorganisms found on the skin. (2014, December 15). Derm Net. Retrieved 2015, October 3. [Web]

11. Montross, S. (n.d.). Extremely Cold Environments. Science Education Resource Center at Carleton College. Retrieved 2015, October 3. [Web]

12. Reporter, D. (2013, October 14). 'I never wash my Levi's': Tommy Hilfiger thinks 'it's crazy' to throw jeans in the laundry after every wear. Daily Mail. Retrieved 2015, October 8. [Web]

13. Splenda | The Too-Good-to-Be-True Product Hall of Fame | TIME.com. (n.d.). Retrieved 2016, April 19. [Web]

14. Turnbull, P. (1966, January 15). *Bacillus*. Medical Microbiology. 4th Edition. Retrieved 2015, October 8. [Web]

15. Zielinski, S. (2011, November 7). The Myth of the Frozen Jeans. Smithsonian. Retrieved 2015, October 3. [Web]

ABOUT THE AUTHORS

Pictured: Kathryn Smith, Ben Huxley, and Seana Thompson (left to right). Not pictured: Mentor Dr. Craig Cary, Professor of Biological Sciences and Director of DNA Sequencing at University of Waikato.

This year in Biotech our team has gained an immeasurable amount of valuable experience. We made our fair share of mistakes, from flooding the autoclave to washing our jeans (which would normally be a good thing), and what we learned from these experiences is priceless. Most notably we learned the importance of communication with each other, our mentor, and our teacher. Although that was our hardest lesson to learn, it is undoubtedly one of the most important. We also learned to deal with changes in schedules and were able to adjust to the roadblocks that appeared along the semester. The importance of being able to work well with others and planning out the steps in research were stressed in this class and we improved our skills in these areas as the year progressed. The knowledge and the fun that we've had this year make us confident that we can be successful in college and in our futures. Thank you for an amazing year.

Effects of blue light on *Danio rerio* (zebrafish) embryo morphology and behavior

Sarah K Childs, Katelynne T Wilkins, Veronica S Postolski and Shawndra L Fordham

Department of Science, Principles of Experimental Design in Biotechnology, Rock Canyon High School, Highlands Ranch, Colorado, USA

In this investigation, we tested the effects of blue light on the development and the morphology of *Danio rerio* embryos by exposing them to different combinations of blue light and white light. The control group had exposure to white light during the day and had an absence of light exposure during the night. The treatment groups consisted of exposure to constant blue light, white light during the day followed by blue light at night, and blue light during the day followed by the absence of light during the night. The effects of the light exposure were measured by taking pictures and videos of the *D. rerio* embryos once a day. Observations of the development of the *D. rerio* were also documented including the time the *D. rerio* embryos emerged from their chorion. A z-test was conducted to determine the similarity between the treatment and control groups. The *D. rerio* embryos from the treatment group exposed to constant blue light were not as developed as the *D. rerio* embryos from control group. This group was three to five developmental stages behind the control group and also had a delayed process of breaking out of their chorions compared to the control group. At most, 20% of the *D.rerio* embryos had emerged from their chorion by the third day post-fertilization whereas 100% of the control group had. The treatment group exposed to constant white light developed and broke out of their chorion at a rate similar to the control group. These results indicate that the blue light ultimately affected the morphology and development of the *D. rerio* embryos.

In 1946, the first computer was introduced; it was the Electronic Numerical Integrator And Computer (EMIAC) which took up an entire room. Now, 70 years later, almost every person has a hand held electronic device connecting them to a world of communication and social media. Studies have shown that 60% of the population spend more than 6 hours a day in front of a digital device,[3] and it continues to rise. The light that is emitted from this technology is referenced as blue light. Because of the rise in technology use, the effects of blue light exposure has peaked scientists' interest. Scientists have determined that blue light is responsible for various biological transformations: increased chances of neoplastic diseases, eyestrain, diabetes, heart disease, and obesity and disruption of the circadian cycle (**Fig. 1**).[3]

The most recent research involving blue light exposure has shown that the light produced by technology, such as phones or computers, results in a reduction in the levels of melatonin and hypocretin in the human brain. Hypocretin, a neuropeptide, regulates wakefulness, arousal, and appetite; melatonin is a hormone in the brain that regulates drowsiness.[3] These are critical regulators of the circadian cycle. Similar to humans, *D. rerio* embryos have both hypocretin and melatonin.[1] When light exposure initiates the circadian cycle, the levels of melatonin decreases to 25%. In dark conditions there is a 100-125% increase of melatonin production in adult D. rerio.[1] Light exposure is fundamental to the circadian cycle because of its regulatory effects on melatonin and hypocretin production.

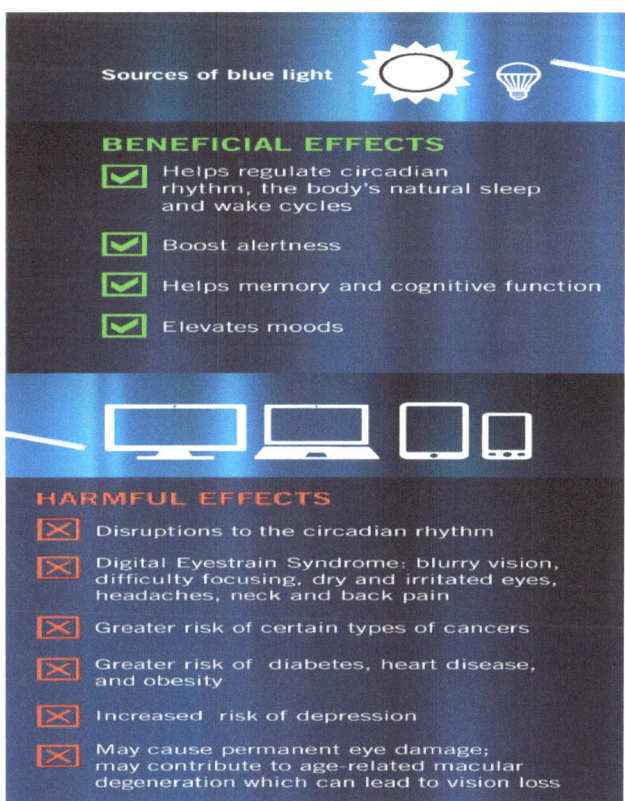

Figure 1: The figure lists positive and negative effects of blue light. Along the top are various sources of blue light that people encounter on a daily basis.[3]

There are some speculations as to why blue light is affecting the body. Blue light waves can be beneficial during the day because they suppress the release of melatonin, increasing attention levels, reaction times, and improving overall happiness. However, at night, blue light is disruptive to the natural release of melatonin in the brain.[3] The effects of melatonin on cancer have been researched by Harvard. According to Cutando, López-Valverde, Arias-Santiago, Vicente, and Diego (2012), the level of melatonin regulates and alters the physiology and biology of cells.[4] The change in the amount of melatonin in the pineal glands correlates to immunosenescence and neoplastic disorders, which are tumor creating diseases. In the study, it was reported that healthy levels of melatonin can inhibit tumor growth in humans.[4] Due to the alteration of the physiology and biology of cells caused by blue light previously noted, the research aimed to study the effects of blue light on the morphology and behavior of *D. rerio* embryos. In the research experiment, it was suspected that more exposure to blue light would cause the *D. rerio* embryos to exhibit increased twitching. Also, the *D. rerio* embryos exposed to more blue light would be less developed and have smaller somite lengths.

The types of light used in this experiment were varying in time periods of exposure to blue light and white light. The wavelength defines the differences between sunlight, blue light, and white light. The wavelength from the sun includes all visible light, ultraviolet and infrared light. The white light used in the trials had the highest wavelengths at 540 nm and 620 nm; however, white light contains the entire spectrum of light. The blue light used in this experiment is characterized as having between 450 nm to 495 nm wavelengths.

According to the diagram by Ballard, Kimmel, Ullmann, and Schilling (1995), there are eight main periods of embryonic development in *Danio rerio*. The first is the zygote (1A); the cleavage is the next stage (2A-2F); the blastula appears after cleavage (3A-3F); then the gastrula stage occurs (4A-4L); next is segmentation (5A-5N); pharyngula occurs after segmentation (6A-6I), hatching (7A-7B); the last stage is the early larval stage (8A) (**Fig. 2**).[5] Embryonic stages of *Danio rerio* development last for 72 hours. The *Danio rerio* were observed between the segmentation and hatching periods.

METHODS

The lab was set up in the back room of the Rock Canyon Biotechnology Lab. This room was chosen because of the ability to control the light and temperature conditions. The lab setup consisted of 5 shelves. Every treatment group was designated to a particular shelf. The lights were situated in the top-corner of each shelf (**Pic. 1**).

Using the lux meter to determine 500,000 lx, the placement of the petri dish that contained the embryos was determined. A tarp from Home Depot surrounded the shelves to eliminate light contamination between treatments. A Ceramic Environment, Lasko heater was used to maintain a constant temperature of approximately 28.5°C in the backroom.

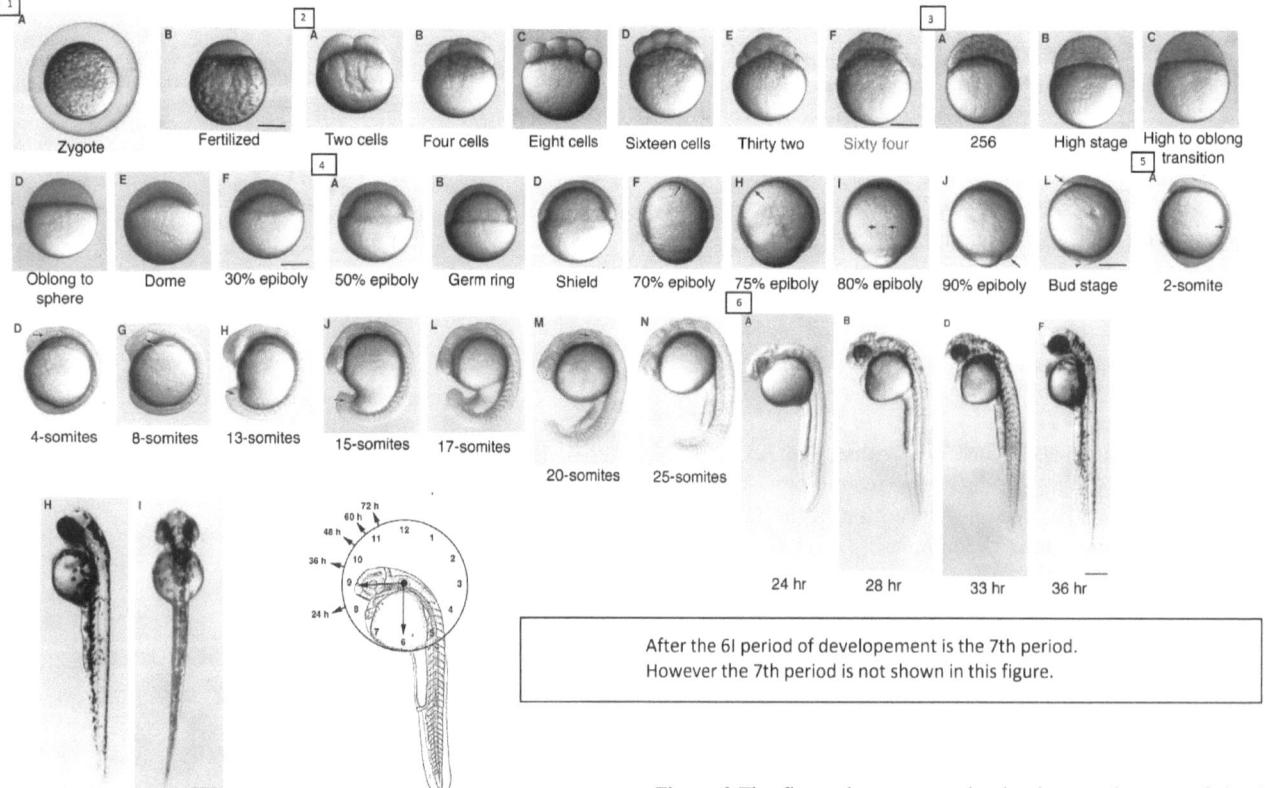

After the 6I period of developement is the 7th period. However the 7th period is not shown in this figure.

Figure 2: The figure demonstrates the developmental stages of the *D. rerio* embryos from the first day of fertilization.[5]

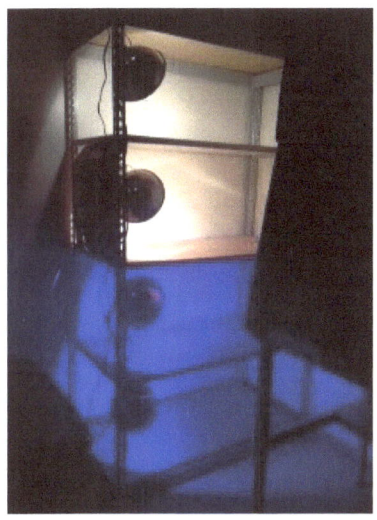

Picture 1: The lab station contained five different shelves for the control group and the treatment groups.

The *D. rerio* embryos and screen media were obtained from the CU Anschutz Medical Campus and were provided by Mellissa Delcont and Tanya Brown, graduate students in the Cell Biology, Stem Cells, and Developmental program at the University of Colorado, Denver. The *D. rerio* embryos were transported back to the Rock Canyon Biotech Lab. By the time they arrived at the lab, the embryos were 10-14 hours post-fertilization, approximately the segmentation stage (**Fig. 2**). The media used for the *D. rerio* embryos consisted of a mixture of aquarium salt in DI H_2O and 0.01 mg/L methylene blue.[6] The fertilized embryos were immediately separated into five small petri dishes. Each dish contained 10 fertilized embryos. Measurements and observations were made on the *D. rerio* embryos once daily for three days.

The light exposure for control and treatment groups are detailed in **Table 1**. The light conditions were changed twice a day, 7:00 A.M. and 3:00 P.M, for three days (**Pic. 2**). The two minute videos and pictures were taken and later used to measure and average twitching per second, when the embryos emerged from their chorions, and their stages of development. Later, a statistical analysis, the z-test, was conducted for all three trials to compare if the population mean for each treatment group to the control group.

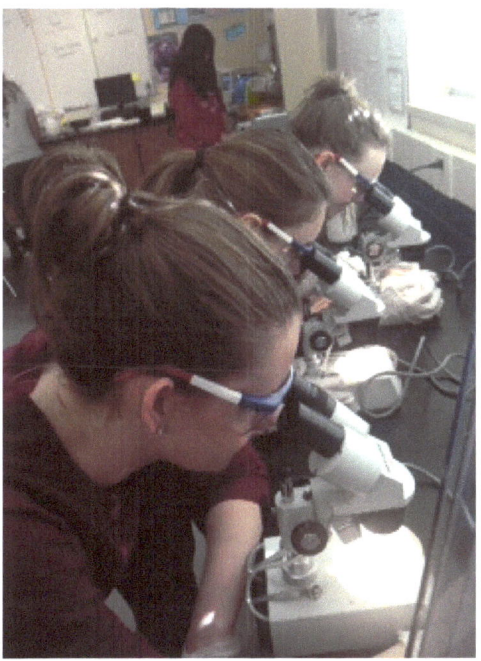

Picture 2: In this picture Veronica, Sarah, and Katelynne are observing the *D. rerio* under the dissecting microscopes.

Light Exposure Type and Duration of Time Each Treatment Group				
Treatment Group	Type of Light	Daytime (hours)	Type of Light	Nighttime (hours)
Control	White Light	7	No Light	17
1	White Light	7	White Light	17
2	White Light	7	Blue Light	17
3	Blue Light	7	No Light	17
4	Blue Light	7	Blue Light	17

Table 1: The amount of hours and type of light each treatment group was exposed to in the experiment.

RESULTS

In this study, *D. rerio* embryos were exposed to different amounts of white light and blue light (**Table 1**). The measurements taken from each treatment group include the average twitching per second, when the embryos emerged from their chorions, and the stages of development.

Activity Level

The average twitching per minute was analyzed to determine the activity level of the embryos. To determine this, the embryos were video recorded for two minutes and the visible twitches were recorded for one minute for each of the individual embryos. The average number of twitches of all the embryos on each day, for each treatment group, were documented (**Table 2**).

Activity Levels				
Treatment Group	Trial 1 (twitches/min)	Trial 2 (twitches/min)	Trial 3 (twitches/min)	Average (twitches/min)
Control	1	5.46	0	2.2
1	0	3	.3	3.0
2	0	5.75	2.8	2.85
3	1.7	0	3.6	1.77
4	0	0	0	0

Table 2: The table documents the average number of twitches for each treatment group in each trial as well as the overall average. This is how we measured activity level in the embryos.

Our data showed no significant correlation between the type of light received by each treatment group and the average times the *D. rerio* twitched per minute (**Table 2**). A statistic analysis was not performed on the data.

Chorion

In all of the trials, all embryos started in their chorions on the first day of observation. By the third day, all the embryos in the control group, treatment group 1 and treatment group 2 had all emerged from their chorions (**Graph 1**). In treatment group 3, an average of 5.667 embryos and an average of 0.1 embryos in treatment group 4 emerged from their chorions throughout the three trials. Based on a z-test, there is no statistically significant difference between the control trial and treatment groups 1 and 2. There is, however, a significant statistical difference between the control group, and treatment groups 3 and 4 with a p-value is less than 0.0001. Only 1 in 10 of the *D. rerio* embryos in treatment group 4, 1 emerged from their chorions while under constant blue light exposure; however, when exposed to ambient light, while recording data on day 3, the embryos began to emerge from the chorions (**Pic 3**).

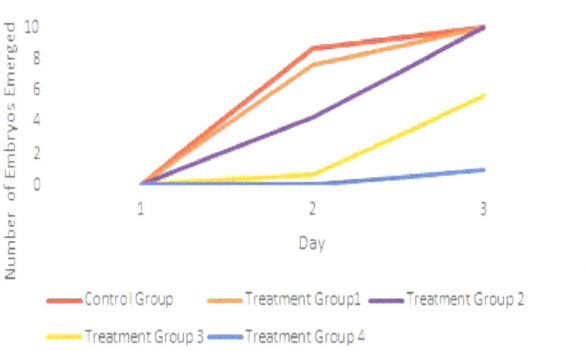

Graph 1: This graph shows the average number of *D.rerio* embryos emerged from their chorion in each trial throughout the experiment.

After fifteen minutes of exposure to ambient light in the lab, an average of 7.33 of the embryos had emerged from their chorions. The embryos broke out of the chorions more rapidly under ambient light than constant blue light.

Picture 3: In this picture, a *D. rerio* is emerging from its chorion. The embryo is in the 6F stage of development.

Developmental Stages

Each treatment group also experienced different rates of development. The *D. rerio* embryos in each treatment group from trials 2 and 3 were averaged together . The embryos in the first trial were not averaged with the second and third trials, because the embryos started at a different period of development. In trial two and three, the average beginning stage of development for the embryos in each

treatment group were 5J and 5L (**Pic. 4a**). By the third day, the *D. rerio* embryos in the control group and treatment group 2 had developed to stage 7A, and treatment group 1 developed to stage 7B (**Pic. 4b**). However, in treatment group 3 and 4, the embryos had developed to the 6D stage of development, which is a 12 hours difference in development from the control group (**Pic. 4c**).

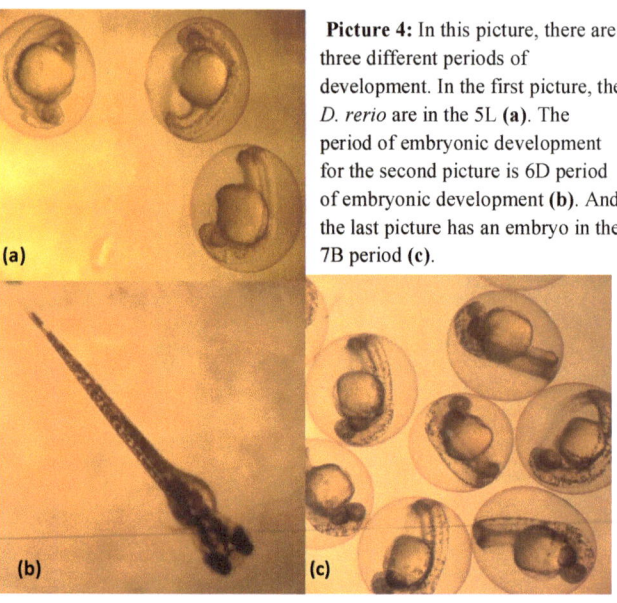

Picture 4: In this picture, there are three different periods of development. In the first picture, the *D. rerio* are in the 5L (**a**). The period of embryonic development for the second picture is 6D period of embryonic development (**b**). And the last picture has an embryo in the 7B period (**c**).

In the control trial, treatment group 1, and treatment group 2 developed faster than the *D. rerio* embryos in treatment groups 3 and 4 (**Graph2**). The results from trial one displayed a similar trend, the only difference is that the *D. rerio* embryos started and ended further along in development (**Graph 3**). From the data collected in all of the trials, the embryos in treatment group 4 were less developed than the embryos in the control group.

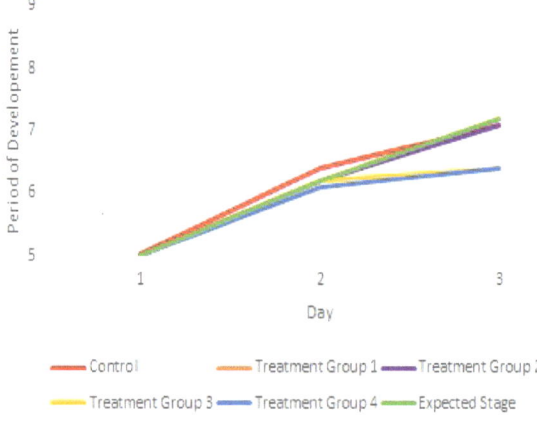

Graph 2: The average period of development that each embryo reached. The periods of development are based on **figure 2.**

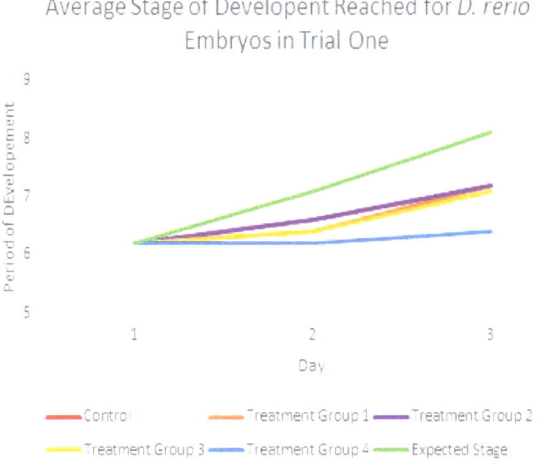

Graph 3: The graph represents the average period of development for the *D. rerio* embryos in trial one. The embryos started at the 6B stage of development.

DISCUSSION

In this investigation the effects of blue light on overall morphology and behavior of common model organism, *D. rerio*, were tested. Data was quantified by determining the stage of development of the *D.rerio* and observing the average twitches per minute throughout the three day developmental period. It was expected that embryos exposed to constant white light, treatment group 1, and constant blue light, treatment group 4, would be different from the control the group (**Table 1**). However, there was no correlation between the light exposure and twitches/minute. The twitching ranged from 0 to 5.75 twitches per minute and were randomly distributed throughout the three trials for the duration of the experiment.

Throughout the experiment, the *D.rerio* in treatment group 4 emerged from their chorion at an earlier stage in development in comparison to the control group (**Graph 1**). *D.rerio* in treatment group 3, blue light followed by no light, also emerged from their chorion later but developed at a more average rate in comparison to treatment group 4. Treatment group 2, white light followed by blue light, emerged from their chorions later but developed at a more average rate in comparison to treatment groups 3 and 4. Treatment group 1, exposed to constant white light, developed similarly to the control group. This was unexpected, because it was previously hypothesised that constant white light would stimulate the *D. rerio* embryos leading to a less developed embryo. According to prior research *D.rerio* embryos begin to break out of their chorions 42 hours post-fertilization.[5] In a study conducted by Kimmel, et al. The sporadic hatching of *D. rerio* embryos, on the third day post-fertilization, is not a reliable indicator of development.[5] In this experiment the *D.rerio* did not hatch sporadically. The results of treatment group 4, exposure to constant blue light, demonstrated that 20% of the embryos hatched at 72 hours post-fertilization, compared to the 100% of the embryos hatched in the control group. This was statistically significant with the p-value < .001. When treatment group 4 was exposed to white light for 15 minutes, the embryos began to emerge from their chorions. This happened consistently during all three trials. This indicates that it is not the excess exposure to the blue light, 450nm - 495 nm, that caused the *D.rerio* embryos to miss important developmental milestones, but the absence of the entire visible light spectrum.

Treatment group 4 demonstrated differences in development in comparison to the control group. It was between three to five developmental stages behind the control group. For all three trials, the developmental data showed that treatment group 1 and treatment group 2 developed similarly to the control group. Even though statistical analysis was not performed for the different stages of development, there was a notable difference between the stages of development within the trials. Throughout all the trials, treatment group 4 was developmentally delayed.

One source of error is that the sample size was too small. Having a larger sample size would increase the accuracy of the results. In future experiments the sample should be increased to 50 embryos per petri plate in order to support the validity of our claim.

Since this experiment did not examine why blue light effects *D. rerio* embryos, the next step would be to treat a group of *D. rerio* embryos without light. This would not only determine the effects of the absence of light but it would also determine if the variation between the developmental stages and the time it took to break out of the chorion was the result of the exposure blue light wavelengths or due to the absence of other wavelengths of light.

ACKNOWLEDGMENTS

We would like to give a special thank you to Sheri Bryant for funding our project and our mentors, Mellissa Delcont, Jessica Spiltoir, Tanya Brown, and Andrew Weems for providing us with their expertise and *D. rerio* embryos. Thank you to Bryan Winkelman for designing our website and for the feedback throughout the whole process. We would like to acknowledge Wendy Lerolland and Heidi Childs for their help with editing the journal article. Also, we would like to thank Tom Dillon for his feedback and support with the design and implementation of our project, and David Ferguson for allowing us to use his spectrophotometer. We would also like to thank Amy Hacker and Susanne Petri for allowing us to use their laboratory space, the Petterle family for lending us their heater, and Jason Dunkle for helping to analyze our data. Thank you Monika Postolski for being generous with your time. Lastly, we acknowledge Rock Canyon High School for providing us with laboratory space and equipment, and Douglas County School District for the Innovation and Perkins Grant. Without the help of these people, this project would never of been possible and we are grateful that they have allowed us to pursue this opportunity.

REFERENCES

1. Appelbaum, L., Wang, G., Maro, G., Mori, R., Torin, A., Martin W. & Mourrain, P. (2009). Sleep–wake regulation and hypocretin–melatonin interaction in zebrafish. *Proceeding of the National Academy of Sciences*, 106(51): 21942–21947.

2. ARAC guidelines: guidelines for use of zebrafish in the NIH intramural research program. (2011). Bethesda, MD; National institute of Health. Bethesda, MD. Retrieved 2015, October 5. [Web]

3. Blue light has a dark side. (2012). *Harvard Health*. 2015, Retrieved October 5. [Web]

4. Cutando, A., López-Valverde, A., Arias-Santiago, S., Vicente, J., & Diego, R. (2012). Role of melatonin in cancer treatment. *International Journal of Cancer Research*, 32 (7), 2747-2753.

5. Kimmel, C., Ballard, W., Kimmel, S., Ullmann, B., & Schilling, T. (1995). Stages of Embryonic Development of the Zebrafish. *National Center for Biotechnology Information*, 203(3):253 - 310.

6. Martinez, M. (2013). Zebrafish Embryo Care, Maintenance and Commonly Used Experimental Approaches. Icahn School of Medicine. Retrieved 2015 October 27. [Web]

ABOUT THE AUTHORS

Pictured: This picture is of Sarah, Katelynne, and Veronica with their mentors, Andrew Weems, Melissa Delcont, Jessica Spiltoir, graduate students with the University of Colorado, Denver. Not pictured: Mentor Tanya Brown also with UC Denver Anschutz. This research couldn't have been done without them.

This class through which we performed this research is a new attribute to Rock Canyon High School. Each group member is a senior at Rock Canyon, and plans on graduating and moving onto an in-state college in 2016. Being part of the first year to take Biotech II, was a great opportunity and we are all so glad we got this opportunity. This class enabled us to experience scientific research at a new level. It exposed us to the outside world while forcing us to utilize our knowledge and skills that are not normally something included in the high school curriculum.

Outside of this research, we all learned how to interact with researchers and professionals, learn challenging topics and develop advanced skills, and find ways to manage difficult situations. These are skills that we will take with us as we move forward throughout our lives. We are prepared for our futures in a way that other students are not and we feel more comfortable moving into college with the skills we have now obtained from this class.

Assaying the effects of coffee on paralysis in Alzheimer's disease model *Caenorhabditis elegans*

Alexa L Vanderhill, Taylor G Hendrickson, and Shawndra L Fordham

Department of Science, Principles of Experimental Design in Biotechnology, Rock Canyon High School, Highlands Ranch, Colorado, USA

For our investigation, we analyzed the effect of coffee on the rate of paralysis in model organism *Caenorhabditis elegans* that had been genetically modified to contain the human Alzheimer's Disease gene, pCL12. Over their lifespan, the *C. elegans* were exposed to coffee and control treatments. For the coffee treatment, we infused the NGM-lite agar with caffeinated Starbucks House Blend coffee and seeded the plates with OP50 *Escherichia coli*. Then we created a synchronized population on both the control and coffee treatments. The number of paralyzed and dead *C. elegans* were recorded every day over a 12 day period. As our research came to a close, we found a statistically significant decrease in paralysis throughout the days of adulthood for *C. elegans* on the coffee plates as compared to the control. By day 12, 100% of *C. elegans* were paralyzed on the control plate, while only 76% were paralyzed when treated with coffee. A 2-proportional z-test, resulted in a p-value of 3×10^{-6} demonstrating throughout their entire life cycle. Strong evidence that when *C. elegans* are exposed to coffee, the rate of paralysis due to Aβ plaque formation declines.

A chronic neurodegenerative disease, Alzheimer's Disease (AD), is suspected to be caused primarily by the accumulation of insoluble β-amyloid peptide (Aβ) between neurons in addition to neurofibrillary tangles of tau protein. Such aggregation of Aβ is what results in the plaque formation associated with AD in humans. Aβ protein aggregation between neurons is also what inhibits communication between synaptic gaps in neurons, resulting in cell death. This neurodegeneration causes the human brain to lose a significant mass of brain tissue, severely inhibiting human functioning, mainly memory and sometimes the ability to communicate with others. AD is a late onset disease which is associated with progressive symptoms such as impairments in cognition and memory. According to The Alzheimer's Association, AD is the 6th leading cause of death in the U.S., with 5.3 million people in America alone being affected.[1] The Alzheimer's Association also claims that of the top ten causes of death, it is the only one that cannot be prevented, cured or slowed. The symptoms and progression of AD vary between various individuals. This same idea is also true when researching the disease in model organisms.

Caenorhabditis elegans were first introduced as a model system by Sydney Brenner in 1963. *C. elegans* are microscopic nematode ringworms that are found worldwide (**Pic. 1**). These nematode roundworm model organisms reproduce quickly, are inexpensive, very small (1mm in length), have short life cycles of approximately 23 days, self fertilize (as they are hermaphrodites), and have a completely mapped cell lineage.[2] Because of this, they are useful organisms to be used for studying human genetics and for studying age related diseases such as Alzheimer's.

For our experiment, we used *C. elegans* strain CL2006, obtained from Dr. Link of the University of Colorado Boulder. These nematodes were co-injected with a

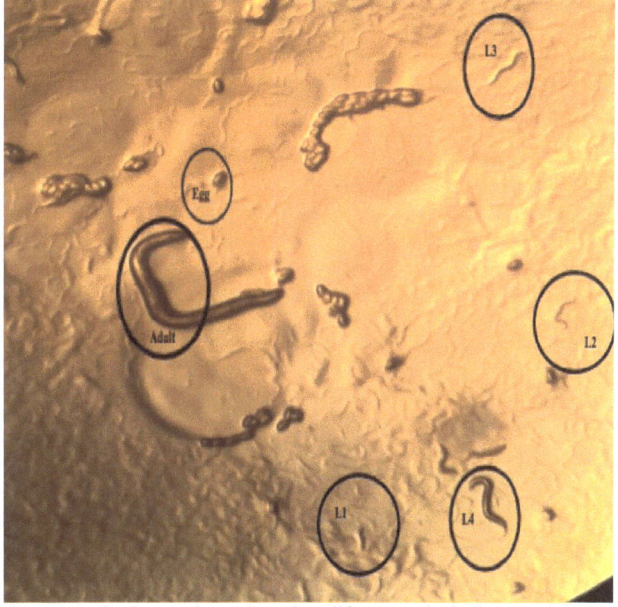

Picture 1: *C. elegans* shown in various life stages, including an adult, L1, L2, L3, and a multitude of eggs

transgene that includes the human Alzheimer's gene pCL12 (P*unc-54*:: SP::Aβ 1–42)[5] as well as a mutant worm gene which results in a rolling phenotype (**Pic. 2**).

Within the model organisms, this transgene causes the expression of Aβ as well as resulting in the subsequent paralysis of the worms. We were able to expose the effects coffee has on the expression of Aβ as well as the rate of paralysis. In strain CL2006, Aβ accumulates in the constitutive muscle tissue of the *C. elegans*, ultimately resulting in the paralysis of the worms because motor neurons reach a point where they are no longer able to communicate as Aβ accumulation interferes in the synapse of neurotransmitters, inhibiting action potential in the neurons to take place and cause abnormal locomotion.

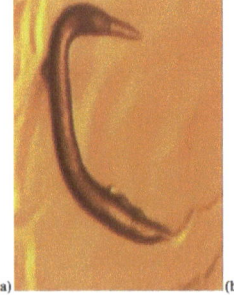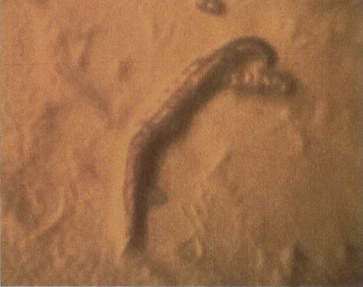

Picture 2: *C. elegan* showing the roller phenotype **(a)**. Instead of moving in a S-like shape, its body is moving in a circle. The paralyzed *C. elegans* are only able to move their heads and tails due to plaque formation **(b)**.

We infused coffee extract into NGM plates and exposed *C. elegans* throughout their lifespan to test the effect it has on Aβ expression and the subsequent effects on paralysis of the *C. elegans* when they are present in the growth media throughout the lifespan of the organisms. In a study by researchers, Vishantie Dostal and Christine Roberts, under the supervision of Dr. Christopher Link of University of Colorado (2010), coffee was shown to reduce the side effects of AD in *C. elegans* Alzheimer model. It was found to be protective against the formation of beta-amyloid plaques. Specifically, coffee extracts have shown to decrease the toxicity of the Aβ peptide, consequently decreasing the onset of paralysis. Caffeine appears to play a significant role in the protection pathways of Aβ peptides. By inducing the activation of SKN-1, the coffee extract enabled a protection pathway to lower the effects of the disease.[3] SKN-1 is the transcription factor that catalyzes the development of mesendodermal tissues, specifically in the digestive system. Postembryonic functions have not been concluded; however, SKN-1 regulates the Phase II detoxification gene during the postembryonic stages caused by the specific mechanisms in the neurons that induce stress. In the brain, coffee acts as an antagonist to the adenosine receptors, which regulate tissue function. These receptors deal directly with inflammatory systems and have now become the target for many therapies in degenerative diseases such as AD.[6]

METHODS

We assayed the rate of paralysis in an AD model of *C. elegans* with lifelong exposure to coffee as compared to the control. We monitored the rate of paralysis and death of *C. elegans* over a 12 day period. We used *C. elegans* strain CL2006, transformed with the Aβ transgene pCL12, provided by Dr. Christopher Link at the University of Colorado Boulder, in order to test how coffee affects toxicity associated with Alzheimer's disease (AD).

We prepared the coffee using protocols originally used in the Pallanck Lab.[7] We added 18.4g of Starbucks House Blend caffeinated ground coffee beans into 100 ml of distilled water and boiled it for 30 minutes. We then used filter paper to remove all solid particles **(Pic. 3)**.

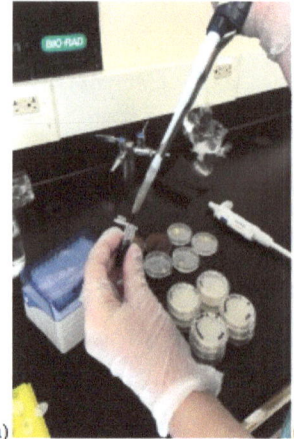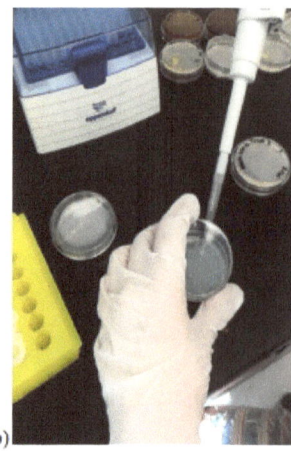

Picture 3: The steps we took to placing coffee on our plates. Beginning with micropipetting 750 µL of coffee from our boiled coffee **(a)** to our microcentrifuge tubes **(b)**. Then we placed 750 µL of coffee onto our 60 mm petri plates.

To determine the dose, we tested doses of 500 µL and 1 mL of Starbucks House Blend coffee. In order to determine dosing for the experiment we observed the transformed worms. We were be able to determine the effects of each dosage of coffee on the *C. elegans* to determine the dosage of the chemicals that were to be used throughout our trials. We determined that 750 µL of 0.184 g/mL coffee was a sufficient amount to diffuse onto the trial plates because it was the highest concentration we could use while keeping the *C. elegans* alive.

In order to test the effect coffee had on paralysis, we diffused coffee onto 60 mm NGM agar plates purchased from IPM Scientific. We began by infusing the plates with 750 µL of coffee, allowing them to diffuse into the media for 24 hours at 22°C. We then added *Escherichia coli* strain OP50 to all the plates, both with and without coffee, and stored them at 22°C for 24 hours. We completed two separate treatments, one with 750 µL of coffee diffused into agar, and the other with plain agar for our control. We prepared 60 mm NGM petri plates in which the two treatments were divided onto two plates for each treatment respectively as well as adding the approximate amount of OP50 bacteria.

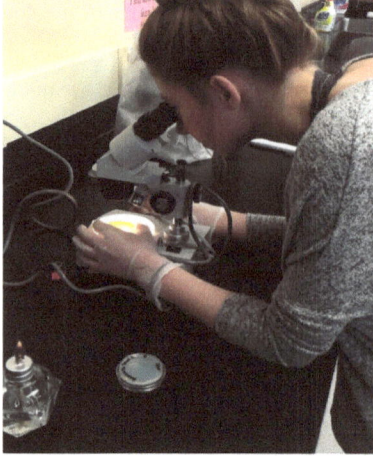

Picture 4: Taylor picking *C. elegans* onto each plate so that we could begin a synchronized population.

Each plate had 10 adult worms placed onto its surface for two hours where they laid approximately 50 eggs total **(Pic. 4)**. After two hours we removed the adults. Each day in the afternoon, for 12 days, we scored the total number of *C. elegans* and counted the number of paralyzed and dead adult worms on each plate. All plates within every treatment were incubated at a constant temperature of 16°C. We completed two trials of two plates for each treatment.

RESULTS

To assay the rate of paralysis the total amount of dead, paralyzed, and alive adult *C. elegans* were counted (**Table 1**). Strictly considering all paralyzed *C. elegans* to be able to move only their heads. The percentage of paralyzed *C. elegans* was in fact less on the coffee plates than the control plates.

Number and Percentage of *C. elegans* Paralyzed Over Adulthood						
Treatment	Day 7	Day 8	Day 9	Day 10	Day 11	Day 12
Control	0.54	31.54	46.54	46.54	53.54	54.54
	0%	57.40%	85.20%	85.20%	98.10%	100%
Coffee	0.38	6.38	13.38	14.38	18.38	28.38
	0%	15.80%	34.20%	36.80%	47.40%	73.60%

Table 1: Each day has a count for the total number of *C. elegans* paralyzed over the total on each coffee and control plate. The percentages of the number of paralyzed adult *C. elegans* out of the total number of *C. elegans* for each day were derived directly from the proportions found and are displayed above.

In addition, using the percentage of the amount of *C. elegans* paralyzed each day, we created an ogive (**Graph 1**). Using this graph, we can clearly assess the rate in which the adults become paralyzed. Looking specifically at day twelve, the coffee plates had 73.6% paralysis amongst all adult *C. elegans,* whereas the control concluded with 100% paralysis. The standard error was calculated to create an interval, which enabled us to find a range in which we were 95% confident the true probability of all samples lies within. Using this method, we compared the control to the treatment and determined the results to be statistically significant with a p value of 3×10^{-6}. This provides more evidence that the average amount of paralyzed *C. elegans* in the presence of coffee is statistically much lower than that of the control.

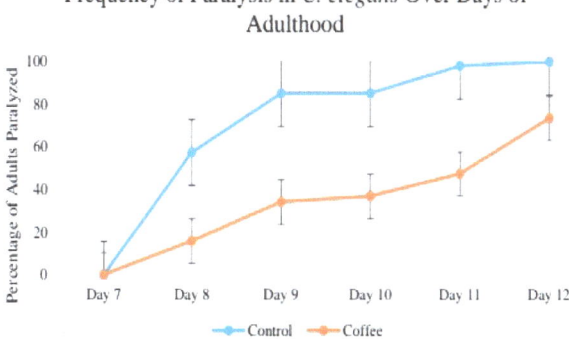

Graph 1: This graph displays the rate of paralysis for both the control and coffee, which was calculated from the data in the previous table. 95% confidence intervals were found for each day's data. The lines indicate the interval in which the true mean for the expected data for each day will be captured 95% of all samples.

We tested the significance for each day by running two proportion z-tests to compare the control trial to the coffee trial. This testing was to conclude whether the data

collected was statistically significant ($\propto =. 01$). It was determined that coffee does in fact have a statistically significant effect on the rate of paralysis due to the calculated $p \geq 0.000003$, which was lower than our set alpha of 0.01. Upon analyzing the data, we found that the results from each day are approximately four to five standard deviations away from the mean. In other words, these results have almost a zero percent chance of happening randomly. In fact, the highest probability of this occurring randomly is 3×10^{-6} (**Table 2**).

Comparision of Control and Coffee Statistical Analysis						
	Day 7	Day 8	Day 9	Day 10	Day 11	Day 12
Z-Score	0	4	5.02	4.794	5.714	3.993
p̂	0	0.402	0.641	0.652	0.772	0.891
P value	0	0.000003	0.0000002	0.0000008	0	0.000003

Table 2: The following table displays a two proportion z-test, which is comparing the percentage of obtaining that result randomly for each day between both coffee and control trials. In addition to percent, the z-score also indicates how many standard deviations away from the expected mean, the results are.

DISCUSSION

Alzheimer's disease is a detrimental disease that impacts the lives of thousands of people. In our experiment, we exposed model organism, *Caenorhabditis elegans*, which had been genetically modified to express the human Alzheimer's disease gene pCL12, to coffee extract infused in the NGM agar plates in order to determine if this lifelong exposure impacts the resulting rate of paralysis. We hypothesized that coffee would lower the rate of paralysis based on previous research using a different strain of *C. elegans.* We have evidence to conclude that coffee does in fact prolong the onset of paralysis, thus confirming our hypothesis.

We found a significant variation in the data between the control, with 100% paralyzed on day 12, and the coffee treatment, with 73.6% paralyzed at day 12. We ran a two-proportion z-test and determined the significance between this data to be 0.000003, meaning the probability of this occurring by random chance is very slim. By running this test, we were able to determine that there is a significant difference in the coffee treatment and control, meaning the coffee had a positive effect on slowing the rate of paralysis. The amount of paralyzed *C. elegans* for both treatments was significantly different on all days of adulthood. In addition to our research, a previous study conducted by Dostal, Roberts, and Link (2010) found similar results. In this experiment, researchers measured the protection coffee had on neural pathways, which further indicates the effects coffee has on Alzheimer's Disease. These researchers used strain CL4176 of *C. elegans*. However, their control was 100% paralyzed after 26 hours, whereas 60% were paralyzed on the coffee plate after 30 hours. We confirmed their conclusions with a different strain of C. elegans in order to determine that coffee reduces the rate of paralysis on the different strain as well.[3] This research is extremely relevant to the real world since we applied different treatments to animal models containing the human Alzheimer's gene. Within our data we saw that coffee does have a significant effect on slowing the rate of paralysis in *C. elegans* containing this gene.

A few errors occurred throughout our experiment. First, there was an error in following the protocol, when picking from our first coffee and control plates to our second plates, resulting in a loss of half of adult *C. elegans* on both these plates. Due to this error, we had to allow the *C. elegans* on one coffee and one control plate to grow to adulthood. These two plates were not synchronized because we concluded the control plate before the coffee plate. This resulted in not having a control plate to use as comparison for the coffee plate on its day 11 and day 12. Another error was miscalculations in the identification of paralyzed versus non-paralyzed *C. elegans*, as several times they appeared to be paralyzed and later moved. We tried to avoid this error by tapping the plates prior to scoring. Lastly, errors could have occurred due to an inconsistency of time that we checked the *C. elegans* for paralysis. Although we did count paralysis every day, it was not always at the same time. The next steps for this experiment would be to use Green Fluorescent Protein (GFP) to measure the amount of plaque build-up between *C. elegans* that have been treated with coffee and the control. Using a microscope and software that can measure the levels of fluorescence, a future experiment could determine the difference between amount of plaque build-up between treatments on a per *C. elegan* basis. A research team could also include measuring how long coffee prolongs the onset of paralysis. In addition, researchers could test the effects other chemicals, including, wine, chocolate, antioxidants, and certain supplements have on paralysis in the same strain of *C. elegans*.

ACKNOWLEDGMENTS

First, we would like to thank everyone who contributed funding for our project, including Orlando Martinez, Tom Bogard, Rob Burkholder, Nicholas Laatsch, Kristen Schurr, Lori Dishneau, Mark Graffiti, and Margaret Koperny. We would also like to acknowledge Dr. Christopher Link of University of Colorado at Boulder. He provided the CL2006 strain of *C. elegans* and mentorship in order to make this project possible.

In addition, we would like to thank Bryan Winkelman for providing an outlet to find research related to our project as well as our class website so that our family, friends, and other people interested in our work could follow our studies. We would also like to thank Amy Hacker and Susanne Petri for allowing us to use their classroom space and for providing support and advice for our project. We would also like to thank Dr. Jason Dunkle as well as Matthew Gracey for assisting us with working out the statistics for our data analysis. We appreciate Andrew Abner and Rock Canyon High School for allowing us to take such an innovative class so that we can pursue this project, and for providing laboratory space and equipment, as well as Douglas County School District for the Innovation and Perkins Grant to fund research grade laboratory equipment to make our project possible.

REFERENCES

1. Latest Alzheimer's Facts and Figures (2013, Sept. 17) . *Alzheimer's Association*. Retrieved 2015, Sept. 24. [Web]
2. Corsi, A., Wightman, B., & Chalfie, M., (2015, June 18). A Transparent Window into Biology: A Primer on *Caenorhabditis elegans*. *GENETICS, 200(2)*, 387-407. Retrieved 2015, Sept. 24. [Web]
3. Dostal, V., Roberts, C., & Link, C. (2010, August 25). Genetic Mechanisms of Coffee Extract Protection in a *Caenorhabditis elegans* Model of β-Amyloid Peptide Toxicity. *Genetics (Impact Factor: 5.96), 186(3)*, 857-866. Retrieved 2015, Oct. 7. [Web]
4. Hasselmo, M. (2006, Sept. 29). The Role of Acetylcholine in Learning and Memory. *Current Opinion in Neurobiology, 16(6)*, 710-715. Retrieved 2015, Oct 7. [Web]
5. Lublin A., & Link, C., (2012, March 10). Alzheimer's Disease Drug Discovery: In-vivo Screening Using *C. Elegans* as a Model for β-amyloid Peptide-induced Toxicity. *Drug Discovery Today Technologies, 10(1)*, 115-119. Retrieved 2015, Sept. 24. [Web]
6. Nehlig, A., Daval, J., & Debry, G.,. (1992, August 17). Caffeine and the central nervous system: mechanisms of action, biochemical, metabolic, and psychostimulant effects. *Brain Research Reviews, 17(2)*, 139-170. Retrieved 2015, Oct. 7. [Web]
7. Trinh, K., Andrews, L., Krause, J., Hanak, T., Lee, D., Gelb, M., & Pallanc,. L. (2010, April 21). Decaffeinated Coffee and Nicotine-Free Tobacco Provide Neuroprotection in *Drosophila* Models of Parkinson's Disease through an NRF2-Dependent Mechanism. *The Journal of Neuroscience, 30(16)*, 5525-5532. Retrieved 2015, Oct. 7. [Web]

ABOUT THE AUTHORS

Pictured: Alexa and Taylor (left to right). Not pictured, our mentor Dr. Christopher Link with the University of Colorado at Boulder.

We learned a plethora of things from performing this experiment including how to work with model organism *C. elegans*. More importantly, however, we learned how to build an experiment up from the bottom and what it takes to work in a research lab. This class has helped us expand our knowledge in not only biotechnology, but along business lines as well. In creating and designing an experiment we encounters many failures and setbacks along way. We have learned to persevere in a way that no other class has ever taught us. We also learned a lot about teamwork and communication as we worked together throughout this year. This opportunity will continue to impact our lives in the future.

Next year, we both will be attending college and I (Alexa) will be studying biology, and I (Taylor) will be studying molecular, cellular, and developmental biology. We have high hopes that the lab skills we have acquired over the past year will help us secure future internships in research laboratories.

The effects of Metformin on *Danio rerio* (zebrafish) embryos

Gayathri Gude, Maia M Bransom and Shawndra L Fordham

Department of Science, Principles of Experimental Design in Biotechnology, Rock Canyon High School, Highlands Ranch, Colorado, USA

A commonly used diabetic drug called metformin has recently entered into clinical trials as a potential cancer treatment drug. Using metformin as a cancer treatment drug, would require a much higher dose than what is now currently used to treat diabetes. The essential question that now arises is what the side effects of higher doses are. The main goal of our study was to expose transgenic *Danio rerio* (zebrafish) embryos to increasing amounts of metformin. These zebrafish, part of a transgenic line called "olig2:egfp, *tg(olig2:EGFP"* caused the cells in the embryos to glow when observed under a fluorescent microscope. We examined the effects on the oligodendrocyte migration along the spinal cord in the region of the yolk sac extension. Since metformin inhibits the oxidative phosphorylation and mTOR pathways, which are critical for cell development and normal cellular function, we hypothesized the metformin would result in a lower migration of oligodendrocytes cells along the spinal cord in this region. To test our hypothesis, we exposed zebrafish embryos to 0mM, 5mM, 10mM, and 20mM concentrations of metformin. Various experimental errors occurred, but when looking at the average numbers of migration and mortality, high concentrations of metformin played a negative role on the embryos. In terms of mortality, more embryos died as they were exposed to increasing concentrations of metformin. In addition to mortality, we noticed that the oligodendrocyte migration was substantially higher in the control group that was not exposed to metformin, as compared to those exposed to metformin. We concluded that increased amounts of metformin results in detrimental effects to development, including an increase in mortality and a decrease in oligodendrocyte migration.

In our experiment, we looked at a diabetic drug, metformin, and its specific effect on zebrafish embryos when given at higher doses. Metformin is often the first drug given to patients with type II diabetes due to its ability to control high blood sugar.[2] Although known for its safety, metformin has been linked to lactic acidosis, the buildup of lactic acid in the blood.[6] Women who take metformin while pregnant have not displayed any negative side effects.[9] In recent studies, metformin has shown the ability to inhibit cancer cell growth through various mechanisms regarding its association with mTOR pathway, such as regulating specificity protein transcription factors in pancreatic cancer cells and tumors and the activation of adenosine monophosphate activated protein kinase, also known as AMPK, to inhibit the growth of breast cancer cells.[3,8] Metformin is currently in phase III of clinical trials regarding its effectiveness with fighting cancer cells. Because the dosage of metformin used to prevent the growth of cancer cells is notably higher than that used to help diabetic patients, it is relevant to the experiment we will be conducting because we will be looking at how the increasing dosages of metformin affect the migration of the oligodendrocyte cells and mortality rate in the embryos.

Currently, many clinical trials are being done to test metformin as an anticancer drug. Dr. Niang, a researcher in the Department of Investigational Cancer Therapeutics of the University of Texas's MD Anderson Cancer Center, has analyzed several scenarios of metformin's potential anticancer activities. In a study done by the Department of Breast Medical Oncology, it was analyzed that in women taking metformin for diabetes in addition to going through chemotherapy, their pathological response was 24% higher than those who had not been given metformin. The University of Texas's MD Anderson Cancer Center currently has six ongoing trials involving the use of metformin as an anti-cancer treatment drug. Metformin, in these clinical trials, is being tested for maximum efficacy by being combined with a number of different therapies. These include chemotherapy, radiation therapy, and targeted therapy.[11]

One pathway affected by metformin is Oxidative Phosphorylation. This is a metabolic pathway where the mitochondria within cells use nutrients to create the molecules that supply energy for metabolism, through the creation of adenosine triphosphate, or ATP.[4] Oxidative phosphorylation is the final stage of the cellular respiration process; all of the NADH and FADH2 produced in glycolysis and the Krebs cycle is used to make ATP, through oxidative phosphorylation (**Pic. 1**).

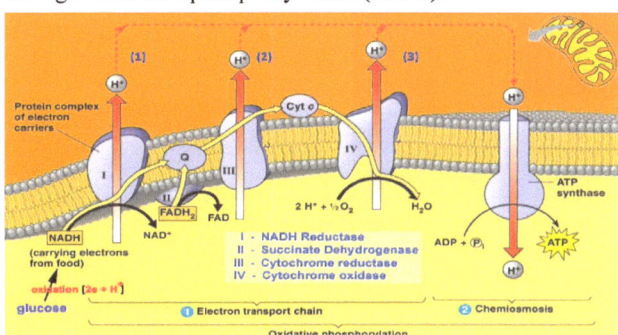

Picture 1: OP Pathway. This figure shows all of the NADH and FADH2 produced in glycolysis and the Krebs cycle that is used to make ATP, through oxidative phosphorylation.

This process takes place at the mitochondrial membrane of the cell. When this process does not happen properly, there is a loss in cell function due to the lack of ATP generation for energy. Cell death can also occur.

The other pathway impacted by metformin is the Mechanistic Target of Rapamycin (mTOR) pathway, which signals the activated protein kinase to regulate cell functions, such as cell growth, protein synthesis, and transcription.[8] The mTOR pathway is particularly targeted within cancer research because there have been mutations found in cancer cells that activate mTOR, which then cause the cancer cell to grow and reproduce.[1]

In this experiment, we looked at oligodendrocyte migration, when exposed to increasing concentrations of metformin. Oligodendrocytes are a type of glial cell, whose main job is to provide for support of the axons in the central nervous system. Glial cells maintain homeostasis and form the myelin sheath, the fatty white substance around axons that is essential for the function of the nervous system. Oligodendrocytes are only found in the central nervous system around the brain and spinal cord.[10] Oligodendrocytes were originally found to be produced in the neural tube; but recent research shows that these could also be produced in some areas of the forebrain and embryonic spinal cord.[12]

We will be specifically counting the number of oligodendrocyte cells that migrate to the neural tube, in the region of the yolk sac extension (**Pic. 2**), and closely observing how the number of cells that migrate is impacted as the dose of metformin increases.

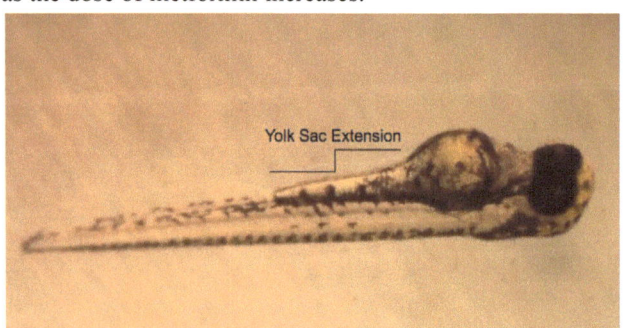

Picture 2: This photo shows where we will be taking count of the Oligodendrocyte cells, along the Yolk Sac Extension.

The embryos are from a transgenic line called "olig2:egfp,*tg(olig2:EGFP"* which causes the oligodendrocytes to fluoresce with the GFP protein, and are able to be seen using fluorescent technology (**Pic. 3**).

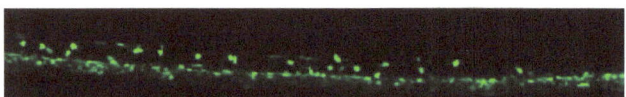

Picture 3: This shows an embryo yolk sac extension (the head is on the left, the dorsal is the upper part, and the ventral is the lower part). This is the exact area of the zebrafish embryo we will be using to take count of oligodendrocyte migration.

METHODS

We started this investigation by setting up an environment, where the zebrafish would be kept at a temperature of 87 °F (**Pic. 4**). Prior to beginning our experiment, we conducted a test with different doses of metformin to determine what the three doses of metformin to use in the experiment. We did this to see which doses are too high and become lethal to the embryos. As a result, we decided on the doses 5mM, 10mM, and 20mM of metformin to add to zebrafish embryo media, which was obtained from Anschutz Medical Campus at CU Denver.

Picture 4: This picture shows the area where we set the temperature to 87 °F. The area directly under the light was the hottest area, therefore we set our fish on the opposite side, where the fish were at an approximate temperature of 87 °F.

Preparation For Experiment

Metformin was obtained in pill form and ground up (**pic. 5**), then weighed to determine the proper amounts for each dosage. The powders were then mixed into 35 mL of zebrafish embryo media.

We then obtained a plate of approximately 90 zebrafish embryos, all from the transgenic line olig 2:EGFP.

After sorting the fish into plates of fertilized and unfertilized embryos (**Pic. 6**) at nine hours post fertilization, we

Picture 5: This picture was taken as we ground the Metformin pill up into the powder to then dissolve into the zebrafish media.

dechorionated the 60 embryos (**Pic. 7**) and ran three trials simultaneously.

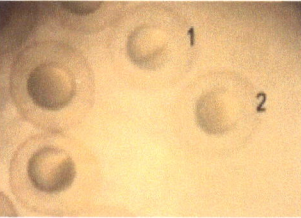

Picture 6: This picture was taken of a plate of zebrafish embryos at 15 hours post fertilization. This is an example of a fertilized embryo (**1**) and an unfertilized embryo (**2**). Note the presence of differentiation, which occurs in the embryo labeled one.

The process of dechorionation was critical to our experiment because in order for the embryos to absorb the drug, the chorion must be removed. Using a pair of fine tweezers and a sharp end probe, we poked the chorion on one side without touching the body of the embryo and twist the probe so that the chorion would wind around it, while simultaneously holding the embryo with the forceps.

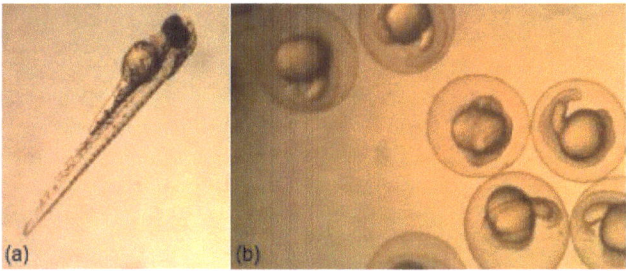

Picture 7: The picture to the left (a) shows an embryo at three days post fertilization that was dechorionated 48 hours post fertilization. The picture to the right (b) was taken of embryos ready to dechorionate at 48 hours post fertilization.

Experiment

Each trial consisted of exposing five zebrafish embryos to 0mM, 5mM, 10mM, and 20mM concentrations of metformin. The trials lasted for three days with mortality rate noted each day. The embryos were housed in a six well plate, with three embryos in each well. After completing the trials, the embryos were transported to Anschutz Medical Campus at CU Denver, where we used the fluorescent microscope to photograph and observe the glowing oligodendrocytes within the embryos **(Pic. 8)**.

Using the photographs, we were able to count the oligodendrocytes that had migrated to the spinal cord in the region of the yolk sac extension **(Pic. 9)**.

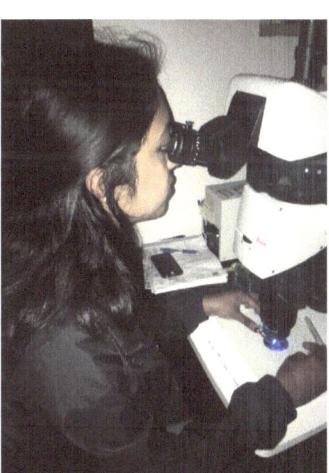

Picture 8: Gayathri is looking at the oligodendrocyte migration of the zebrafish embryos under the florescent scope at UC Denver.

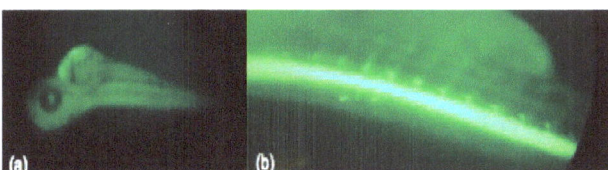

Picture 9: The picture to the left (a) shows the fish under the florescent scope, prior to zooming in to count the cells. The picture to the right (b) shows the same fish in picture 13 under the florescent scope, but was taken after zooming in to count the cells. This is taken of the region along the embryos' yolk sac extension. The glowing dots are actually the oligodendrocyte cells that have migrated.

RESULTS

After analyzing our data, we observed that as the embryos are placed in higher concentrations of metformin, there is an increase in mortality. In trial one, there was a random pattern of mortality. Throughout all the trials, all fish died in the 20mM of metformin. The trend supported by the data is that as the concentrations of metformin increases, so does mortality. While the data seems to be randomly distributed, there was a significant difference between the control with an average mortality of 13.33% and the 20 mM metformin with 100% mortality. However, due to inconsistencies in the 5 mM and 10 mM metformin treatments, the overall trend cannot be observed **(Graph 1)**.

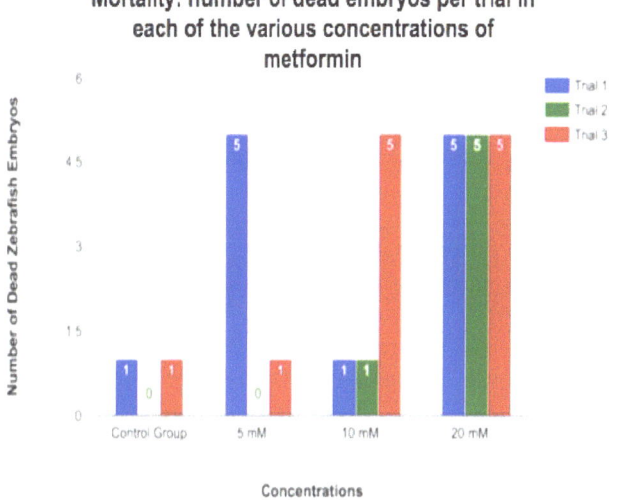

Graph 1: This graph shows how many zebrafish died when exposed to each of the concentrations of Metformin. Each of the different colors represent a different trial, as shown by the key.

In all trials we observed that as the concentration of metformin increased, the oligodendrocyte migration decreased **(Graph 2)**. The embryos in the control had the greatest migration with an average of 8 oligodendrocyte cells migrating along the neural tube. Oligodendrocyte migration could not be measured for the 20 mM treatment group, due to the 100% mortality rate observed. In trial 2, there was less oligodendrocyte migration in the control group in comparison to the 5 mM treatment group. An average of 5 oligodendrocyte cells migrated within the control group, which is less than the 3 oligodendrocyte cells that migrated within the 5mM metformin treatment. This was not expected, in comparison to the general trend of increasing mortality when exposed to increased concentrations of metformin.

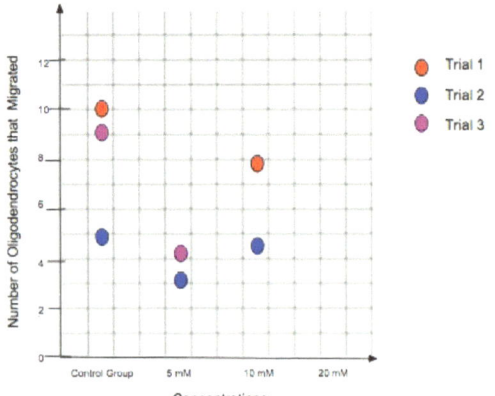

Average Oligodendrocyte Migration at Various Concentrations of Metformin

Graph 2: This graph shows the average oligodendrocyte migration per concentration in each trial. During trial one, all fish in the 5 mM plate died; therefore, no oligodendrocyte migration could be measured. All fish in the 10 mM plate of trial three died and all fish throughout all trials in the 20 mM plates died.

DISCUSSION

In this experiment, we were investigating how varying amounts of a potential cancer treatment drug, metformin, would impact the neurological development in zebrafish embryos. To test the development of the embryo, we measured oligodendrocyte migration to the neural tube. A decrease in this migration would affect the nervous system, as the myelin sheath would not be formed to protect the nerve fibers. This would eventually lead to spinal cord failure.[12]

Initially, we hypothesized that high concentrations of metformin would correlate with negative side effects, as measured by mortality rate and oligodendrocyte migration, seen in the embryos. This experiment ultimately showed a general trend that increasing doses of metformin result in negative impacts on zebrafish embryos. However, because of low sample sizes, problems with experimental design, inconsistent results, and a lack of the ability to perform statistical analysis, no conclusions can be drawn based on our data. One of the side effects we expected to see in the embryos was a decrease in the number of oligodendrocytes migrating to the neural tube. The oligodendrocyte migration decreased as the concentration of metformin increased. While mortality rates were highest in the greatest, 20mM, concentration of metformin.

The greater number of oligodendrocytes found along the neural tube in embryos in the control plate, as compared to those exposed to metformin, raises the question of how the increasing concentrations of metformin, impacts normal oligodendrocyte migrational patterns. Since oligodendrocyte cell migration is critical to the developmental myelination process and creation of the central nervous system, which both directly relate to the mTOR pathway, there is a possibility that it is the impact to mTOR that resulted in this decrease. More research needs to be done to determine if this is the case.

Throughout our experiment, there were many experimental errors that resulted in our inability to draw a strong conclusion. One of these errors centered around the removal of the chorions. Removing the chorion is very difficult and our mortality rates could be due to damage to the embryos and not drug exposure. In addition to this, embryos are very sensitive to temperature fluctuations and we struggled throughout the experiment to maintain a stable temperature. Another significant error occurred only in our very first trial, which had significant fluctuation in mortality. We kept our metformin solutions refrigerated, but we did not warm them to room temperature prior to adding the embryos. This caused a temperature shock, resulting in higher mortality, than normal, as well. This was fixed for the remainder of the trials and all of the solutions were stored at room temperature as were the embryo plates.

In future studies we recommend increasing the sample size as well as exposing the embryos to Rapamycin along with metformin. Rapamycin directly impacts the Oxidative Phosphorylation pathway. These trials would have serve as a comparison to the metformin trials, to observe if it was the mTOR pathway or the Oxidative Phosphorylation pathway that was impacting oligodendrocyte migration.

ACKNOWLEDGMENTS

We would like to thank Tanya Brown and Alex Liggett with University of Colorado, Denver for their time and patience and guiding us through our project, as well as providing us zebrafish embryos. We would like to also thank the following people for kindly donating money towards our experiment: Tom Bogard, Rob Burkholder, Nicholas Laatsch, Kristen Schurr, Lori Dishneau, and Mark Grafitti. We appreciate Dr. Matthew Fordham with CA Mountain View Animal Hospital for very kindly providing the metformin for our experiment. Thank you to Bryan Winkelman for organizing our websites and guiding us on our blogs as well as supporting us throughout our research. We want to thank Amy Hacker and Suzanne Petri for sharing their laboratory space and equipment, as well as Rock Canyon High School for providing the laboratory space and equipment. We would further like to thank the Douglas County School District for the Innovation and Perkins Grant funding that provided research grade laboratory equipment. We also would like to thank Tom Dillon for his feedback on the design of our project. Thank you to Uma Venkitanarayanan for guiding us through any questions we had while writing the results and discussion of the paper.

REFERENCES

1. Advani, S. H. (2010, August) Targeting mTOR pathway: a new concept in cancer therapy. Retrieved 2016, February 22. [Web].
2. Alexandria, M. (2013) Effectiveness and safety of Metformin in diabetic patients with kidney problems: American Diabetes Association®.Retrieved 2016, February 28[Web].
3. Ali. (2013) Metformin: Uses, Dosage, Side Effects. Drugs.com. Retrieved 2016, February 28 [Web].
4. Berg J. M. (2012, August) Oxidative Phosphorylation. Retrieved February 28, 2016, [Web].
5. Brown T. (2014). *Oligodendrocyte Under Fluorescent Technology.* [digital image]. Retrieved from Denver, Colorado: UC Denver's Anschutz Medical Campus.
6. Ekstrom N. (2013). Effectiveness and Safety of metformin in Diabetic Patients With Kidney Problems. American Diabetes Association®. Retrieved 2015, November 11 [Web].
7. Jakubowski (2015) Figure One. Oxidative Phosphorylation. NCBI. Retrieved 2015, November 11 [Web].
8. Kasznicki J., Sliwinska A., & Drzewoski J. (2014). Metformin in cancer prevention and therapy. Retrieved 2015, November 11 [Web].

9. Kumar, P, & Khan K. (2010). Effects of metformin use in pregnant patients with polycystic ovary syndrome. Retrieved 2015 November 21 [Web].
10. Lovell K. (2013). Oligodendrocytes - structure/function. Retrieved 2015, November 21. [Web].
11. Patterson, S. (2014). Metformin may have broad utility in cancer. Retrieved November 2015 23. [Web].
12. Springer S. (2013). Regulation of oligodendrocyte precursor migration during development, in adulthood and in pathology. Retrieved 2015, November 23 [Web].
13. Sottdard B.H. (2006, April). Metformin: Can a Diabetes Drug Help Prevent Cancer? Retrieved 2016, February 25 [Web].
14. Yue W., Yang C., Dipaola R., & Tan X. (2014). Repurposing of Metformin and Aspirin by Targeting AMPK-mTOR and Inflammation for Pancreatic Cancer Prevention and Treatment. Retrieved 2015, December 1 [Web].

ABOUT THE AUTHORS

Pictured: Our mentors, Alex Ligget (left) and Tanya Brown (right) with the University of Colorado, Denver along with Gayathri and Maia (center).

We were so excited and fortunate to be able to experience a second year of biotech. Being able to experiment and conduct our own research is not something that many high school students can say that they have done. Taking this class was a look into the "real world" which is something that not very many high school courses can offer. We were able to experience real labs and lab equipment, and communicate with people who actually work have backgrounds in biotechnology related subjects as well as people working in them currently. The most important lesson we learned by taking this class was that mistakes do not equal failure. As someone who is prone to messing up, the research we came up with came across many obstacles. From our initial experiment being rejected to accidentally almost starting a fire due to a heat lamp malfunction, we faced more frustration than in any of our other classes. However, with encouragement from classmates, mentors, and especially our teacher, we were able to come up with solutions and ways to fix things that were not working. Finishing the research not only gave a sense of accomplishment but also showed us what we were capable of if we worked hard at something. The experiments and lab work we were doing were not skills that average high school students would have. It has provided us with hands on experience that is advanced for high school and even college students. Going into college having had this experience will make us stand out. This class taught us scientific writing, professionalism, teamwork, and collaboration. It also required us to act like real researchers, coming in to school on the weekends and before and after class. Scheduling other classwork, extracurricular activities, and jobs with our experiment was definitely challenging, but we were able to learn about time management and balance. Overall, taking this class was an incredible experience and has been extremely beneficial not only to our futures but also personally. We will always be grateful for this opportunity.

A comparison of the effects of two herbicides, Roundup® and Beyond®, on the mortality, morphology, and cardiac development of *Danio rerio* (zebrafish) embryos

Mallory F Happ, Megan N Happ, Taylor N Kelly, Heidi S Markel and Shawndra L Fordham

Department of Science, Principles of Experimental Design in Biotechnology, Rock Canyon High School, Highlands Ranch, Colorado, USA

The experiment compared the effects of an herbicide commonly used with genetically modified soybeans, Roundup®, and an herbicide used with mutation bred sunflowers, Beyond®, on the cardiac and morphological development of *Danio rerio* (zebrafish) embryos. The experiment was performed using a base concentration of each herbicide, with the same ratio of main ingredient to water (1:320). The embryos were exposed to different concentrations of each herbicide and monitored twice a day for two days. We measured mortality and heart rate, and noted other behavioral and morphological observations. It was found that there was an overall increase in heart rate for embryos exposed to herbicides compared to the controls, with embryos exposed to the highest concentration of Roundup® having a higher average heart rate (164.125 bpm) than embryos exposed to the highest concentration of Beyond® (160.269 bpm) compared to the average control heart rate (154.379 bpm). However, the higher average heart rate was not found to be statistically significant between the two herbicides (p= 0.779 between Beyond® and Roundup®), or between each herbicide and the control (p=0.405 control and Beyond®) (p=0.219 control and Roundup®). A greater mortality rate was recorded for Roundup® in comparison to Beyond® (4.167 to 2.778 death per well). This difference was statistically significant because the p value was 0.0097. The difference between the mortality rate for Beyond® and the control was statistically significant (p=0.057), however the difference between the mortality rate for Roundup® and the control was not (p=0.111). Differences in heart size, pigmentation, and yolk sac development were also noted between the embryos exposed to herbicides and the control, although not directly measured. After measuring heart rate and mortality over the three trials, it appears that Roundup® has greater side effects when compared to Beyond®; however, our data were not statistically significant enough to fully support this statement. Throughout the experiment, we noticed an increase in the rate of development and heart size in embryos exposed to both herbicides at the highest concentrations. We recommend further studies be conducted that directly measure these factors in addition to confirming impacts to heart rate and mortality.

In society today, there has been much controversy over genetically modified (GM) crops versus conventionally grown crops. Recently, Chipotle claimed to be "GMO free" by switching from using a soybean oil, derived from genetically modified soybeans to be resistant to Roundup®, to non-GM BASF Corporation sunflower oil derived from mutation bread sunflowers with Beyond® herbicide. Chipotle claimed the switch was better for multiple reasons such as, "Soybean oil is almost always made from genetically modified soybeans [but] sunflowers, however, have not yet been genetically modified, thus making sunflower oil a great non-GMO alternative" **(Fig. 1).** [3]

They claimed that this makes the use of soybean oil from GM soybeans worse for the environment, as this single-herbicide use leads to the emergence of "superweeds." Chipotle claimed the switch from soybean oil to sunflower oil would combat these issues.

As of today, Chipotle has removed these claims from their website. However, we continued to test Chipotle's claim by comparing the effects of two different herbicides, Roundup® and Beyond® on the development of zebrafish embryos. *Danio rerio* (zebrafish) are common model organisms for studies in developmental biology due to their common genetic and physiological factors with humans. Therefore, we plan to investigate which herbicide, Roundup® or Beyond® has worse developmental effects on zebrafish embryos which are genetically similar to humans.

The BASF Corporation herbicide, Beyond®, is used in combination with sunflowers that have been mutation bred to be resistant/tolerant to imidazolinone herbicides. Both Beyond® and the sunflower hybrids work together to provide maximum weed-control. Roundup®, in combination with Monsanto's genetically engineered Roundup® resistant soybeans, disrupts the shikimic acid pathway which is necessary for plant growth and protein synthesis.[1] In our literature review, there has been no assessment of repeated dose toxicity, genetic toxicity, reproductive toxicity, or teratogenicity, could be found for Beyond® herbicide. In contrast glyphosate is one of the most heavily tested herbicides in the United States, and has been classified by the Environmental Protection Agency as

SCIENTISTS ARE STILL STUDYING THE LONG TERM IMPLICATIONS OF GMOS.

While some studies have shown GMOs to be safe, most of this research was funded by companies that sell GMO seeds and did not evaluate long-term effects. More independent studies are needed.

THE CULTIVATION OF GMOS CAN DAMAGE THE ENVIRONMENT.

Evidence suggests that GMOs engineered to produce pesticides or withstand powerful chemical herbicides damage beneficial insect populations and create herbicide resistant super-weeds.

CHIPOTLE SHOULD BE A PLACE WHERE PEOPLE CAN EAT FOOD MADE WITH NON-GMO INGREDIENTS.

In our quest to serve the best ingredients, we decided to remove the few GMOs in our food so that our customers who choose to avoid them can enjoy eating at Chipotle.

Figure 1: Above are Chipotle's previous claims stating why they moved from GM to non GM products. *(Figure 1, images 1-3 taken as screenshots from Chipotle website (September 2015) from claims made by Chipotle in 2013. Claims have since been removed from the website by Chipotle.)

a group D risk for cancer, indicating that there is no sufficient evidence or data to demonstrate a cancer risk.[2] However, the International Agency for Research on Cancer of the World Health Organization in March 2015 classified Roundup® as "probably carcinogenic to humans," due to "mechanistic evidence" such as observed damage to DNA in human cells as well as the presence of tumors in mice and rats[5]. Researchers found that embryos treated with the glyphosate-based herbicide showed stunted neural crest development, and morphological alterations such as decreased anterior-posterior width.[4] There is a large amount of research on glyphosate based herbicides such as Roundup®, but not on Beyond®, showing the importance of comparing the effects of each and determining if one is more harmful than the other. The findings of these studies further support the importance of research on the effects of all herbicides and how they may impact vertebrates and their development, particularly in humans.

The goal of our experiment was to compare how Beyond® and Roundup® impact the heart rate and morphology in zebrafish embryos. In our experiment we used Beyond® and Roundup® herbicides with three levels of concentrations at 8.01e-5%, 3.47e-4%, and 7.83e-4%. We ran three trials over a five-day period, measuring heart rate, morphology, and general observations twice a day.

METHODS
We conducted our experiment by running three trials of five days each during which we measured the heart rate and mortality rate while also observing morphological development of zebrafish embryos in different concen-trations of Roundup® and Beyond® herbicides.

Preparation
GFP-Oligo2 (transgenic) zebrafish embryos and wild-type

zebrafish embryos were obtained from the University of Colorado Denver Anschutz Medical Campus. Four six-welled petri dishes (Corning® CellBind® Product #3335) were used to house the embryos during the experiment. The herbicides in each well compared Beyond® Clearfield Production System Herbicide to Roundup® Weathermax Herbicide. These products were both diluted to create a ratio of 1 mL of herbicide to 320 mL of tap water in order to create a 0.3125% starting concentration and to best mimic their preparation for agriculture use. The herbicides in these wells were further diluted from their original 0.3125% concentration to 8.01e-5%, 3.47e-4%, and 7.83e-4% for the three trials. The embryos in solutions were kept in a temperature-controlled environment ranging between 26-28.5 °C **(Fig. 2)**.

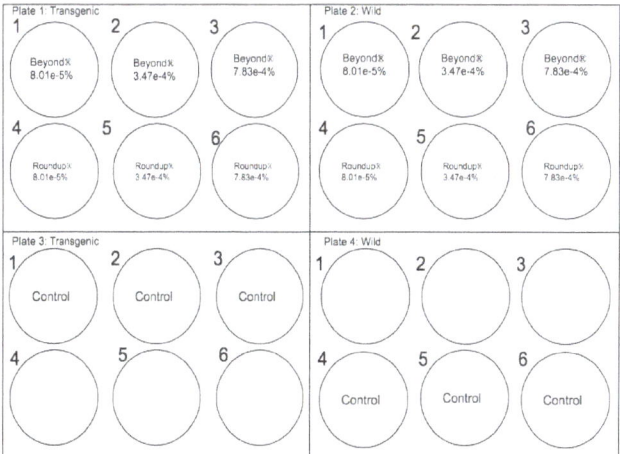

Figure 2: Above a diagram that explains the distributions of % concentrations of herbicides that were used through all three trials.

Trials
First, embryos were separated into fertilized and unfertilized embryos **(Pic. 1)**. Unfertilized embryos were sacrificed. Fertilized embryos were then put in a petri dish and their chorions were removed **(Pic. 2-3)**. After their chorions were removed, 4-5 embryos were placed in each well **(Pic. 4)**. During the trial, we measured heart rate through videos taken every morning and afternoon. Beats per minute were calculated by counting heart rate over a 10 second time. Mortality rate was also measured. Behavior and morphology, including tail movements and pigmentation were observed and recorded from pictures of the embryos.

Data Collection
A t-test was used to analyze for statistical significance for heart rate and mortality. Averages of heart rate and mortality across all three trials were also analyzed. Heart size, development, and morphological observations were collected during the trial and from embryo pictures as qualitative observations.

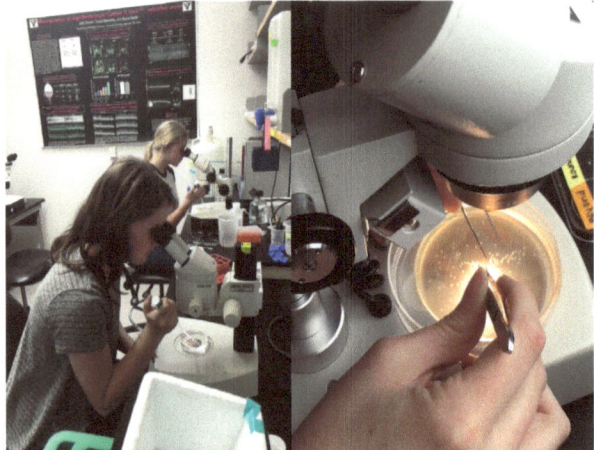

Picture 1: Megan and Heidi sorting fertilized and unfertilized embryos at our mentors' lab.

Picture 2: This picture shows a close up view of the process of breaking chorions.

Picture 3: Taylor and Mallory breaking embryo chorions.

Picture 4: Taylor and Mallory placing embryos in each well.

RESULTS

In addition to general observations throughout each trial period (2 days), the heart rate, development, morphology, and mortality rate were also observed. Due to a lack of significant data for statistical purposes, many of the results are reliant on the visual data provided by the photographic documentation throughout each experiment.

Heart Rate

The average heart rate for embryos exposed to Roundup® was 164.125 bpm while the average for embryos exposed to Beyond® was lower at 160.269 bpm (across all three trials). The difference in heart rate between embryos in Roundup® and Beyond® was not statistically significant t=5.4 (p= 0.779) and neither were statistically significantly different from the controls (p=0.405 t=0.847 Beyond® compared to control) (p=0.219 t=1.465 Roundup® compared to control). The average heart rate in embryos exposed to Roundup® was 1.595 standard deviations when compared to the control average, and the average heart rate of embryos exposed to Beyond® was 0.463 standard deviations when compared to the control **(Graph 1-2)**.

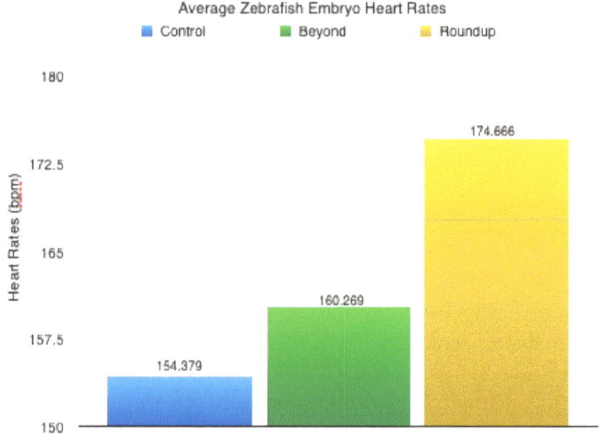

Graph 1: This graph demonstrates the difference in heart rate averages among the control, Beyond®, and Roundup® concentrations across all three trials with all % concentrations. *Due to the large differences in standard deviation standard deviation bars are not displayed on this graph.

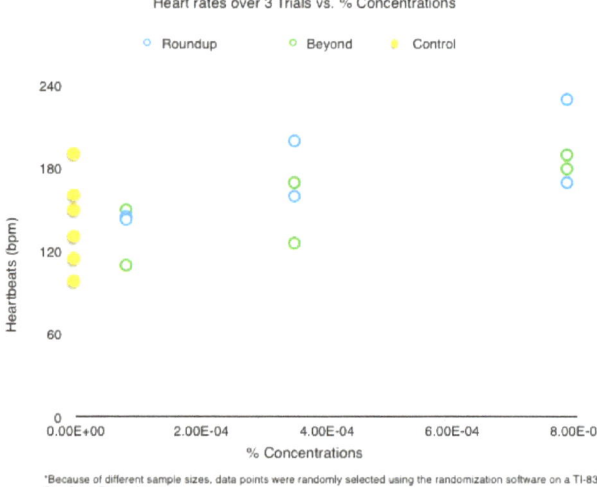

Graph 2: The scatter plot demonstrates that as % concentration increased, heart rate also increased across the three trials and shows Roundup® heart rate to generally be above Beyond® heart rate.

Mortality

Both Beyond® and Roundup® showed increased mortality when compared with the control samples. Control samples had an average mortality rate of 3.533 deaths per well, Beyond® had an average of 2.777 deaths per well and Roundup® had an average of 4.166 deaths per well. A higher overall mortality rate was observed for wells exposed to Roundup® than wells exposed to Beyond® (explaining lack of data points for other parameters for Roundup® wells). A t-test resulted in a p value of 0.0097, showing the higher Roundup® mortality rate is statistically significant when compared to the Beyond® mortality rate **(Graph 3-4)**.

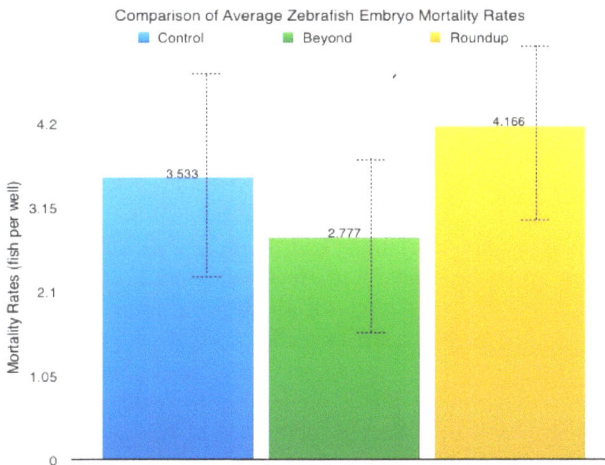

Graph 3: The graph shows the difference between average Beyond® and Roundup® mortality rates as an average all three trials.

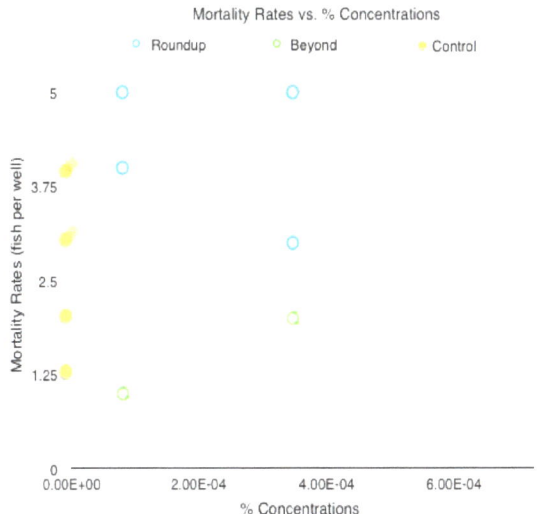

*Because of different sample sizes, data points were randomly selected using the randomization softwa Plus Silver Edition Calculator. Data points with a shadow point show a second data point of that same \ concentration.

Graph 4: This graph shows that the Beyond® mortality rates were consistently lower than Roundup® mortality rates across all three trials. Beyond® mortality rates seemed to increase as the % concentration increased whereas Roundup® did not follow this pattern.

Observations on specific causes of mortality were also noted. For many of the dead embryos in trial 1, the embryos exposed to Roundup® appeared disintegrated, fuzzy, and extremely deteriorated, whereas the dead embryos exposed to Beyond® were severely deformed **(Pic. 5).** Trial 3 also included what appeared to be dead embryos that were underdeveloped and deformed for control, Beyond®, and Roundup® wells. For trial 3, the dead embryos in the control plates also appeared disintegrated.

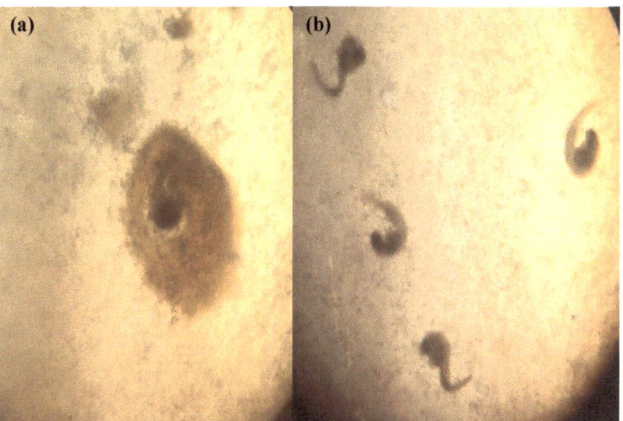

Picture 5: Above is an example of the deaths typical to embryos exposed to Beyond® **(a)** and Roundup® **(b)** during trial 1 of our experiment. The dead embryos exposed to Beyond® appear deformed, whereas the dead embryo exposed to Roundup® appears unrecognizable and disintegrated.

Morphological Observations

Throughout the experimental trials, it was qualitatively observed that embryos with higher herbicide concentrations for both Beyond® and Roundup® seemed to have larger heart sizes than those in wells that were not exposed to herbicides **(Pic. 6).** This observation was noted at the end of both trials 2 and 3. Although this data were unquantifiable with the resources available, the smaller hearts were visually estimated to be approximately 2-3 times smaller than those typical of the previous experimental.

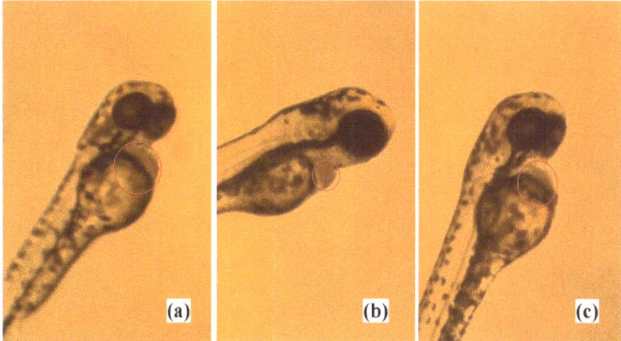

Picture 6: Above are visual data demonstrating the difference in heart size noted during trials 2 and 3. The heart size of embryos exposed to higher concentrations of Beyond® **(a)** and Roundup® **(c)** was greater than the heart size of the control embryos.

The pigmentation characteristic of the body of the developing zebrafish embryo was also observed. An increased concentration of dark pigmentation on the bodies of the embryos were observed as Beyond® concentration increased; however, in trial 2, (after a significant lower in temperature), wells with Beyond® were observed as appearing underdeveloped with little to no spots, appearing almost entirely transparent on the outside **(Pic. 7).**

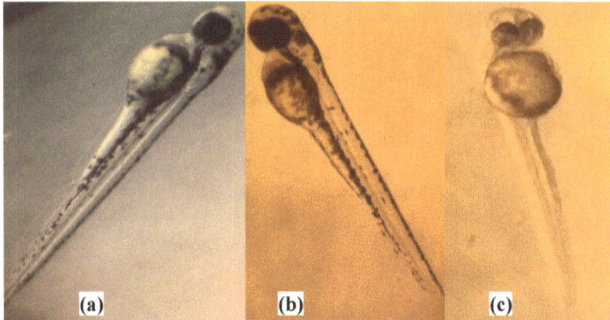

Picture 7: These images show the variation in pigment among embryos exposed to Beyond®. The left most picture (a) shows an embryo exposed to a lower concentration of Beyond® with lighter pigmentation in comparison to the center picture (b) showing an embryo exposed to a higher concentration of Beyond® with darker pigmentation. The right most picture (c) shows an embryo exposed to a middle concentration of Beyond® with little to no pigmentation, demonstrating the variety in pigmentation among embryos exposed to Beyond®.

Finally, overall fish size and shape were noted throughout the trials. Several of the yolk sacs in the surviving Beyond® wells for trial 1 were severely deformed, and it was observed that at higher concentrations of herbicides as in trial 1, the embryos displayed curved bodies and less uniform development **(Pic. 8).** Once the concentration of Roundup® was recalculated (for trial 2 and trial 3), the size and shape of the embryos were more consistent across the control, Roundup®, and Beyond® wells.

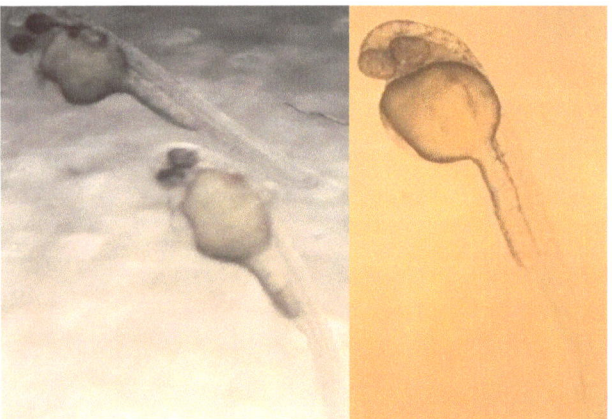

Picture 8: These images demonstrate the deformities in development in embryos exposed to Beyond® during trial 1, primarily showing deformed yolk sacs and curved bodies.

DISCUSSION

Our findings have implications for glyphosate and ammonium salt of imazamox-based herbicides and their effects on the development of vertebrates such as *Danio rerio* (zebrafish) embryos. Specifically, the mortality rate and the heart rate throughout the trials might have revealed the potentially detrimental effects of these two herbicides on living organisms. Additionally, the morphology and qualitative observations throughout all three trials could indicate accelerated or stunted development as well as how these herbicides affected the way that the embryos died.

The goal of our experiment was to compare how Beyond® and Roundup® impact the heart rate and morphology in zebrafish embryos.

Heart Rate and Heart Size

The results showed that across all three trials, both herbicides had an effect on the heart rates and possibly on the size of the embryo's hearts. For all three concentrations of Beyond® herbicide, the heart rates generally increased over time. Additionally, all three concentrations of Roundup® herbicide showed increased heart rate (with a few outliers) across the days of the trials. The embryos exposed to Roundup® had higher heart rates on average (164.125bpm) than the embryos exposed to Beyond® (160.269bpm); however, this was not a statistically significant difference (p=0.779). Outliers with exceptionally low heart rates and small sample sizes in the Roundup® plates could have affected the overall average. Lower mortality, thus a larger sample size, could help to prove statistical significance in the future. The heart rate data collected from embryos exposed to Roundup® were 1.595 standard deviations and the heart rate data collected from embryos exposed to Beyond® were 0.463 standard deviations from the control trials showing embryos exposed to Roundup® had heart rates farther from the controls. In comparison to the averages of the heart rates for both control plates, the plates containing herbicides showed a higher heart rate that either increased or held constant over time. The higher heart rates exhibited by the embryos exposed to herbicides indicate that the herbicides themselves could be causing embryo stress, altering their heart rates.

While observing the heart rate, we were also able to see trends in the heart size of the embryos. Across all trials, it was often difficult to distinguish the heart in the embryos, as they were sometimes extremely small. During trials 1 and 2, the hearts appeared to develop at a slower rate in embryos exposed to either herbicide than observed in the control embryos. However, during trial 3, the embryos exposed to the herbicides appeared to have larger hearts than those observed in the control plates, with separate chambers that could be easily distinguished. As we did not measure heart size quantitatively, no concrete conclusions can be drawn from the varying heart size; however, it is important to note this observed change, which we encourage to be explored in future experiments. With respect to heart size, there were no observed differences between embryos exposed to Beyond® and embryos exposed to Roundup® when each were compared to the control.

Mortality

The results also indicated interesting patterns in mortality rates among the embryos. After evaluating our data for mortality, we were able to see a statistically significant trend in which embryos exposed to Roundup® had a higher mortality rate than those exposed Beyond®. During a three day period, on average 2.777 embryos per well in Beyond® concentrations died and 4.166 embryos died per well in

Roundup® concentrations. There was a statistically significant difference between both of these averages and the control embryos showing that both Roundup® and Beyond® were harmful to the embryos (p=0.057 t=1.465 Beyond® compared to control) (p=0.111 t=1.711 Roundup® compared to control). However, embryos exposed to Roundup® had statistically significant higher mortality rates compared to those exposed to Beyond® (see results, p value= 0.0097). Even though both herbicides showed significant effects on the embryos' mortality rates, the increased mortality among embryos exposed to Roundup® could imply that it is in fact worse for vertebrates, like zebrafish embryos, in comparison to Beyond®. However, further research would be required in order to fully substantiate this claim.

In addition to the mortality rates, the ways that the embryos died varied drastically. Higher concentrations of both herbicides showed embryos that had disintegrated into the solution, and fully intact embryos were difficult to find, if even present. Additionally, dead embryos were observed as being darkened, deformed, or deteriorated around the edges with increased concentration of both herbicides. The cause of death of the embryos could provide insight into the specific effects of the herbicides on the tissues of the embryos themselves, however they could also be due to other external factors such as temperature or stress. Further trials with larger sample sizes and a larger variety of doses would need to be conducted in order to draw reliable conclusions from the cause of death of the embryos, however, it is possible that the herbicide did affect the way the embryos died in our trials.

Morphology and Conclusions

Previously conducted research showed that glyphosate-based herbicides produced teratogenic effects on vertebrates by impairing retinoic acid signaling. After incubating *Xenopus laevis* embryos with 0.02% dilutions of a commercial glyphosate based herbicide, Paganelli et al (2010) concluded that glyphosate-based herbicides have an effect on morphology in *Xenopus laevis* embryos.[4] In our trials, we exposed *Danio rerio* embryos to three different concentrations of glyphosate-based herbicide: $8.01e^{-5}$%, $3.47e^{-4}$%, $7.83e^{-4}$%, as compared to the 0.02% dilution used by Paganelli et al. Paganelli's experiment focused on neurological effects including cephalic markings and neural crest development in addition to morphological deformities. While our research did not specifically look at neurological effects, it did examine morphological deformities consistent with those studied by Paganelli et al. The outcomes of Paganelli et al (2010) and the results of our research may suggest that glyphosate-based herbicides may lead to irregular development and morphological deformities.[4]

Errors and Future Research

During our experiment, a few possible errors arose that could affect the way that our data were interpreted, and the results of our experiment. Throughout all three trials, there were problems with the general environment the fish embryos were being kept in. The temperature of the environment fluctuated throughout the trials, and was maintained in inconsistent ways which could have influenced mortality and caused what were recorded as effects of the herbicides. In addition to issues with temperature, it is also possible that during the earlier trials, more embryos were injured during dechorionation due to unfamiliarity with the technique of dechorionation, possibly influencing our data (specifically mortality rates).

The results of our research were relatively inconclusive due to our experimental errors and small sample size. In future experiments related to our research, more time should be allotted to test a larger sample size of fish, and a better control over variables such as temperature, technical consistency, and timing should be exercised.

ACKNOWLEDGMENTS

First, we would like to thank our mentors, Tanya Brown, Veronica Fregoso, Stephanie Bonnie, Ashley Bourke, and Mellissa Delcont with the University of Colorado Anschutz Medical Campus for their extensive help with our research, including help with our methods, materials, and data analysis. They graciously provided us with zebrafish embryos, shared their advanced lab space and equipment, and taught us new techniques necessary to complete our research in addition to providing us with insight into the field of biological research. Additionally, we would like to thank Amy Hacker for her contributions to our research proposal, for her helpful guidance and assistance with facilitating our experiments, and for her and laboratory space. We would also like to thank Susanne Petri for sharing laboratory space. In addition, we are thankful for the help and writing support of Bryan Winkelman who not only helped with our article, but also helped us document our research through bi-monthly web-posts. We would like to thank Dr. Jason Dunkle for helping us analyze our data. We also would like to recognize Tom Dillon for his feedback and support. We would also like to recognize Rock Canyon High School, including the administration and science department for supporting our research, providing laboratory space and equipment, and allowing us to have the freedom to pursue our research interests. We are also grateful for Douglas County School District's contribution through the Innovation and Perkins Grant funding that has provided us with research grade laboratory equipment. We also greatly appreciate those who helped to fund our research including Tom Bogard, Jean Whittier, Rob Burkholder, Nicholas Laatsch, Kristen Schurr, Lori Dishneau, Jay Kelly, Mark Grafitti, and Margaret Koperny.

REFERENCES

1. Active ingredient fact sheets (2015, September 11). National Pesticide Information Center. Retrieved on 2015, August 30. [Web]
2. Charles, D. (2015, April 30). Why we can't take chipotle's GMO announcement all that seriously. Colorado Public News Radio: The Salt. Retrieved on 2015, August 30. [Web].
3. Chipotle Mexican Grill. (2015). Food With Integrity. Retrieved on 2015, August 30. [Web]
4. Paganelli, A., Gnazzo, V., Acosta, H., López, S. L., Carrasco, A. E. (2010). Glyphosate-based herbicides produce teratogenic effects on vertebrates by impairing retinoic acid signaling. Chem. Res. Toxicol., Vol. 23. doi: 10.1021/tx1001749.
5. Thomas, A. (2015, March 20). IARC monographs volume 112: evaluation of five organophosphate insecticides and herbicides. World Health Organiza World Health Organization: International Agency for Research on Cancer. Retrieved on 2015, August 30. [Web]

ABOUT THE AUTHORS

Pictured: Our team and our mentors, Veronica Fregoso and Stephanie Bonnie, from the University of Colorado Anschutz Medical Campus at our formal research pitch.

Over the year, we not only answered our research question and saw the effects of the two herbicides on zebrafish embryos, but also learned valuable lab and teamwork skills. Because zebrafish embryos are model organisms, we are likely to work with them in future labs and now are very familiar with them. We know their sensitivity to temperature and what their normal and abnormal development looks like. We can now dechorionate embryos in seconds, a process that initially took almost an hour. We've learned how to work with each other on a long-term, high-stakes project and through that have grown as researchers and as a group. This project has enabled us to make valuable connections with our mentors at the University of Colorado Anschutz Medical Campus. They have showed us what research looks like at a graduate level and taught us difficult lab techniques and how to use advanced equipment. More importantly, we got a taste of the research industry. Throughout our entire classroom, we observed the endless possibilities research has to offer. Through conducting our own research, we learned that nothing works out the first time. Things must be changed and fixed over and over again. We learned not to become discouraged when the data did not turn out the way we expected. Our group is confident that after this experience, we are ready to face the challenges of research again in college and further on.

The allergenicity of ovalbumin under heat degeneration

Analissa A Merkle, Mark A Hinkle, Keegan A King, and Shawndra L Fordham

Department of Science, Principles of Experimental Design in Biotechnology, Rock Canyon High School, Highlands Ranch, Colorado, USA

Food allergies are a growing problem globally. We researched whether the method of preparation of eggs, a common food allergen, affects the binding of IgG antibodies to ovalbumin. The first step in this process was to prepare the raw, fried, baked, and autoclaved egg samples. A Bradford test was then run to ensure that the starting concentration of protein present in each sample would be approximately equivalent. Once the concentrations were standardized, an ELISA test was used to compare binding with the control, water, and the four different treatments of eggs. Our results did not show a significant difference in the binding that occurred between the proteins in the different egg preparations and the IgG antibody. The fried egg sample bound the most, as was evident with a mean absorbance of 0.425 Au, and was very close to the amount of binding that occurred in the water control, which had a mean absorbance of 0.334 Au. The raw egg samples bound the least, with a mean absorbance of 0.300 Au, which can be compared to a mean absorbance of 0.328 Au from the autoclaved sample and a mean absorbance of 0.365 Au from the baked egg sample. Since these results were so close, and the 95% confidence intervals for each of the egg samples was either fully contained in the control interval, or a majority of the confidence interval was within the water control, no statistical differences between the treatments were found. Our conclusion is that the cooking methods did not significantly affect the amount of ovalbumin binding that occurred with the IgG antibodies in our egg samples.

Allergic reactions to food affect approximately 365 million people globally, ranging from mild intolerances to life threatening conditions.[5] In many cases, even the most experienced immunologists can be unsure what causes certain foods to be rejected by the body and how to prevent the allergic reaction from occurring. This is often due to the amount of factors that are involved in the immune system, such as IgG, IgE, T-cells, mast cells, and immune complexes, as food is digested. Researchers are beginning to find a direct correlation between the way in which a food is prepared and its allergenic properties. In a study of 108 patients tested for egg allergies, utilizing both raw and cooked eggs, 38 patients reacted to both heated egg whites and raw eggs, 29 patients reacted only to the raw eggs, and the other 41 patients displayed no reaction to eggs whatsoever, indicating that they did not have any egg allergies.[4]

By cooking a food item instead of leaving it in a raw state, proteins have the ability to denature due to the increase in heat. This leads scientists to believe that protein degeneration can keep a human's immune system from recognizing a potential allergen. One of the ways the human body combats biological threats is by using immunoglobulin E (IgE) antibodies; the immune system recognizes antigens on the surface of harmful proteins and binds to them, signaling the body to release histamines. In addition to altering the shape of the protein, heat degeneration affects these antigens on the surface of proteins. If the IgE antibodies are unable to recognize the protein, it is less likely that an allergic reaction will occur.

Though there are many factors that affect protein allergenicity, it is mostly due to the protein's resistance to both digestive enzymes and heat.[1] This means that the change in a protein during cooking can have a direct affect on the allergenicity, depending on that protein's resistance to heat or various cooking processes.

In eggs, the two most common proteins linked with allergies are ovomucoid and ovalbumin.[1] The specific protein tested in this experiment was the protein found in eggs known as ovalbumin, as previous research had shown that ovalbumin is much more influenced by heat than ovomucoid.[2] Researching this protein will suggest possible methods to prepare eggs to reduce its harmful allergenic properties. Interestingly enough, people with allergies to ovalbumin typically do not have allergic reactions when consuming baked products containing eggs.[7]

One way of determining a protein's allergenicity is by using a test known as an ELISA test. This stands for enzyme-linked immunoabsorbent assay, a process where antibodies are attached to the ELISA plate and bind to the protein in question. The solutions are then dyed such that the concentration of protein can be measured using a spectrophotometer. Typically the antibody that would be used to identify ovalbumin is an IgE; when IgE is not able to be used, IgG can be substituted, as it functions in the same way but is a synthetically produced antibody that our lab is able to use. Based on this experiment, we expected that our control (raw egg) would bind the most to the IgG antibody in our ELISA test, thus relating to a stronger allergic reaction due to it not being treated with heat. We predicted that the baked egg sample would have the least amount of binding due to the heat being the greatest, causing the proteins to degenerate the most. The tests were used to reveal if cooking alters the protein ovalbumin in a way that makes the antigen markers less reactive with the IgG antibodies in the ELISA tests.

In addition to our Enzyme Linked Assay we used a Bradford Assay, prior to running the ELISA tests, to ensure that each sample of egg began at the same concentration of egg protein in the solutions. Without this crucial step, much of the data we gathered from the ELISA test would have been influenced by the amount of buffer in our egg samples, instead of the actual binding of IgG and ovalbumin. We used this information to dilute each egg solution to the same concentration before running them in the ELISA test.

METHODS

The egg sample preparation was the first portion of the experiment. We purchased 24 large grade AA eggs from our local King Soopers. Using these store brand eggs, we prepared a solution using raw egg whites that were separated from the yolks and then mixed and shaken with DI water to form a more liquid solution. The solution was then drained through a Melitta White #4 Coffee Filter to ensure an all liquid solution. We then transferred the filtrate to a test tube and stored it at 4°C. The additional eggs were prepared in three different ways: frying, baking, and autoclaving. The first step for each preparation type was isolating the egg whites and was followed by cooking the egg samples. The fried eggs were prepared on a Presto cool touch electric griddle at 300°F for approximately 10 minutes, until cooked thoroughly into a relatively solid form. The baked eggs were prepared in glass bowls sprayed with original Pam Cooking Spray and baked at 350°F in an oven (**Pic. 1**). The eggs were cooked for 20 minutes to ensure they were fully cooked throughout the sample. The autoclaved egg sample was cooked in a Tuttnauer EZ10 autoclave at 250°F for 30 minutes (**Pic. 2**). DI water was added to all three egg preparations and the solids were ground up using a mortar and pestle, filtered through a coffee filter, and then stored at 4°C.

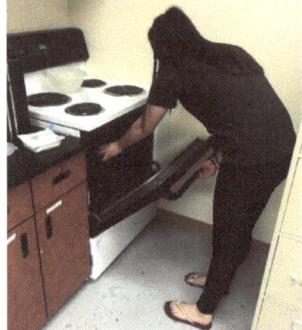 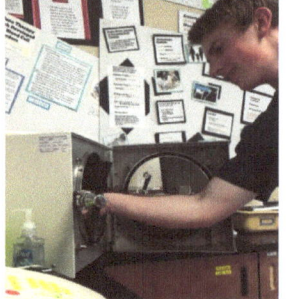

Picture 1: Analissa is putting eggs in the oven to be cooked.   **Picture 2:** Mark is putting eggs into the autoclave.

The next step of this experiment was to perform a Bradford test using 300ul of each sample in order to determine the starting concentration of proteins within each sample type. This test was run using the Quick Start

Bradford Protein Assay from Bio-Rad. The standards were set for the Bradford test which was accomplished by making various solutions of DI water and gamma globulin standards. Solutions of DI water and egg solution were created based on the standards with the substitution of the egg samples instead of gamma globulin. After all the samples were loaded into an ELISA plate, 1X dye reagent was added and the absorbance was tested using an iMark Absorbance Plate Reader from Bio-Rad (**Pic. 3**). Based on the results, the egg samples were brought to the same protein concentration using a standard absorbance of 0.500 Au. This allowed for our egg samples to have the same approximate starting protein concentration.

The final stage in our experiment was to perform an ELISA test using IgG antibodies (**Pic. 4**). The protocols for the experiment were followed according to the Bio-Rad ELISA Immuno Explorer Kit instruction manual using ovalbumin antibody instead of the primary antibody provided by the kit.[3] In addition, the antigen from the kit was replaced with our egg samples and the amount of wash buffer used was lowered from 900 ml to 350 nml. The ELISA plates were loaded into the iMark Absorbance Plate Reader (**Pic. 3**) in order to give us the absorbance.

Picture 3: This is the spectrophotometer that was used to detect the wavelength of light from the ELISA plate.   **Picture 4:** Keegan and Mark are loading the ELISA plates with samples.

The data collected from each trial were combined to analyze the different treatments. Each of the data sets were normally distributed and the mean score and standard deviations of each was used in all of the treatments to build 95% confidence intervals. The confidence interval of each egg sample was compared to the control sample (water) to measure the difference.

RESULTS

Contrary to our predictions, there was little to no difference in the binding that occurred in our samples. The absorbance of each ELISA well reflected the amount of binding that existed between the samples and the IgG antibody. The greater amount of binding that occurred, the higher the absorbance would be, due to light not having the ability to pass through the solution as easily. In our data (**Pictures 5 & 6**), all of the ELISA wells are similar in color, indicating that the data would be expected to have similar absorbance levels. The data from each of the three trials was combined and, due to small sample sizes, a t-test was run to find the sample size (n), mean score (x-bar), and standard deviation

(Sx) of each treatment and the control (**Table 1**). Using this data, a confidence interval of 95% (t* = 2.571) was created for each of the sample types and was compared between the control (water) and each of the treatments (**Graph 1**).

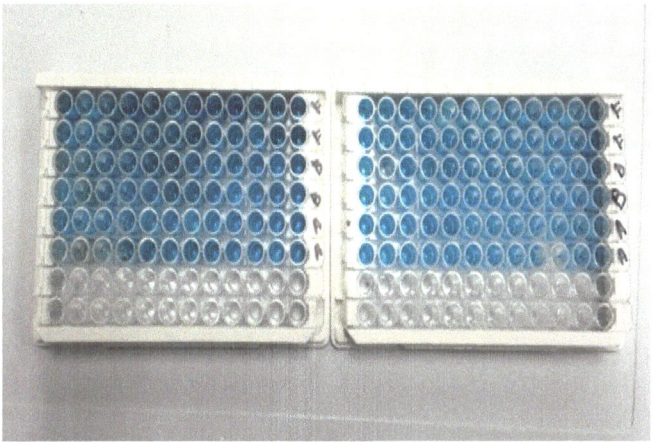

Picture 5: The ELISA runs for Trials 1 and 2 were ran at the same time and are pictured. The plates were formatted the same way. Column 1 contained the raw egg sample, while column 2 contained the control (water) sample. Rows A and B (from column 3 to column 12) contained the fried egg samples. The baked egg sample was in rows C and D (columns 3-12), and the autoclaved egg sample was in rows E and F (columns 3-12).

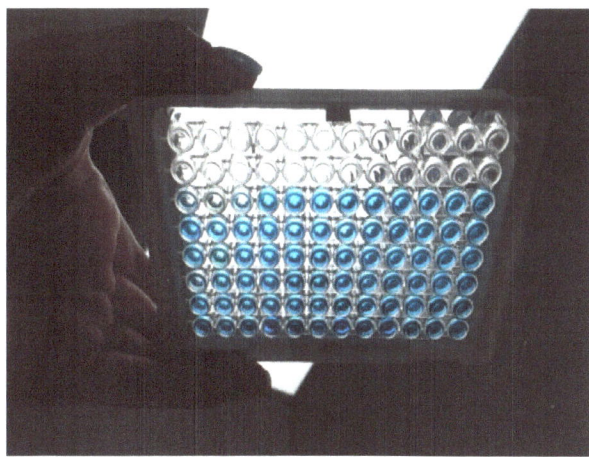

Picture 6: The ELISA plate for Trial 3 (pictured) was formatted the same way as the runs for Trials 1 and 2.

Graph 1: This graph displays the confidence intervals of the mean scores of the absorbance of each egg treatment across all the trials using a t* of 2.571.

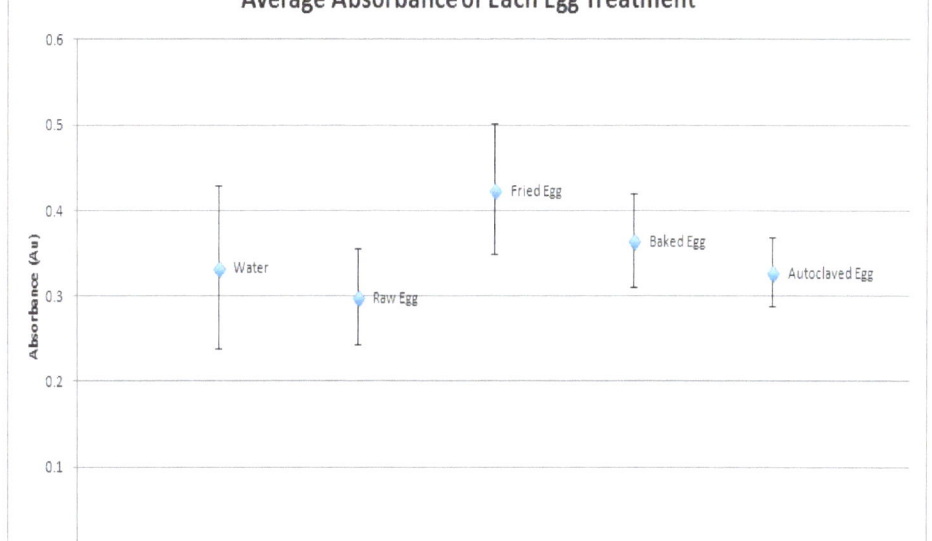

Combined Absorbance Data for Each Treatment			
Treatment	n	x-bar	Sx
Water	18	0.334	0.148
Raw Eggs	18	0.300	0.093
Fried Eggs	60	0.425	0.229
Baked Eggs	60	0.365	0.167
Autoclaved Eggs	60	0.328	0.121

Table 1: Each of the three original trials were averaged to create data for each treatment as a whole. The sample size (n), mean score (x-bar), and standard deviation (Sx) were found for each sample type using a t-test.

By looking at the data presented by the mean scores of the data (**Table 1**), the samples do not seem significantly different. The confidence intervals given in **Graph 1** support this data, as the confidence intervals of each egg treatment either partially or fully overlaps with the control trial of water. This expresses further that the samples were not statistically significant or different.

DISCUSSION

Fully uncovering heat sensitive allergies could be just the beginning of new findings that bring relief to the millions of people who struggle with them. The purpose of our research was to determine which method of preparing eggs would yield the least amount of binding with synthetic antibodies made to imitate real immune cells in a human body. Our team hypothesized that baked eggs would yield the least amount of binding due the high amount of heat that would degenerate its protein more than other cooking methods.

What we found was actually quite surprising and different than what we expected. Our results showed that there is little to no difference in binding between the variety of preparation types. Based on the confidence intervals we gathered from our data, the samples all had very similar absorbances to each other; the differences between them were not great enough to conclusively rate one egg sample above or below any of the others. Due to the fact that the data we recorded, from the spectrophotometer varied greatly within individual sample sets; the confidence intervals they created were large enough to eliminate much of the difference in absorbance we gathered from the means. This is multiplied by the fact that our water samples, the intended negative control, resulted in data that was often similar in value to our egg samples, when, in fact, the absorbance should have been near zero. In trial 1 water samples yielded absorbances of 0.676 Au and 0.318 Au, levels so far apart that they spanned the entire range of our data. This was consistent enough in every trial that when we compared our samples against water, we were unable to statistically differentiate them from each other. If our team had used a sample of pure ovalbumin at a known concentration for our positive control, then much of the cooked egg samples would be usable and differentiable. Unfortunately, the raw egg data could not be considered a positive control, having too large of a confidence interval in the data.

Because of this, no conclusions can be drawn from our data. This is largely due to the fact that our negative control yielded very scattered results, making it nearly impossible to differentiate our cooking methods.

These results do not align with many of the allergy trials we have seen thus far, which typically suggest that cooked eggs are less allergenic than raw eggs.[6] The data we collected contrasts a similar experiment that stated that 50-85% of people with any kind of egg allergy do not show allergic reactions to eggs that were baked.[8] This experiment showed a lower average of binding that occurred within the baked egg sample, whereas our experiment found that the egg samples were too similar to be differentiated. This unusual result could spark more interest on this topic and

hopefully inspire further research on heat sensitive food allergies.

The best way to improve the accuracy of this experiment would be ensuring that the negative control acts as expected so that sample data can be interpreted. In our research specifically, we found that higher amounts of binding occurred for our water control than we expected, which should theoretically not occur as the antibodies should not bind with the water. This most likely resulted from the use of plates that were scratched on the bottom or contained residue that would cause unwanted absorption in the spectrophotometer. To avoid this problem in the future, ELISA plates should be kept in the packaging and away from test materials or rough surfaces to reduce the chance of contamination. To keep the confidence intervals of each sample small, it is imperative that measurements are as accurate as possible, especially when adding materials to the ELISA plate; this should help further secure the accuracy of the results.

The most appropriate next steps for continuing the research conducted in this experiment would be using IgE antibodies in a higher BSL lab, as well as testing other allergenic foods that are suspected to be influenced by heat degeneration. Scientists at Johns Hopkins University School of Medicine have identified milk allergies to be similarly susceptible to degeneration by heat compared to eggs.[6] A similar ELISA test using human antibodies at an offsite lab, measuring milk and egg allergenicity over 5-10 minute time intervals of baking at 350°F, with milk kept in a sealed container to prevent evaporation, would be an effective future study. This experiment could assist in finding the appropriate amount a cooking either of these foods would need to reduce the allergenicity of the food. The precise quantities of materials needed should be confirmed as early as possible to prevent material issues as well as assisting in the planning of the research to create a realistic timeline.

ACKNOWLEDGMENTS

We would like to thank our mentors, Dr. Vijaya Knight and Patricia Merkel, from National Jewish Health for their assistance in the creation of our research question and assistance in our procedural methods. A special thank you also goes to Dr. Damon Tighe for mentoring us throughout our project in regards to our protocols, methods, and the ordering of materials.

We would like to thank the following people for funding our research project: Lillie Aguilar, Tom Bogard, Rob Burkholder, Lori Dishneau, John Hinkle, Margaret Koperny, Nicholas Laatsch, Karen Magner, Thomas Merkle, and Kristen Schurr. We would especially like to thank Adele and Joseph Merkle and Robert Merkle for their financial contribution to our project.

Another thank you goes to Rock Canyon High School, and our principal, Andrew Abner, for providing laboratory space and equipment, along with a special thank you to Amy Hacker and Susanne Petri for graciously sharing lab space with us. Thank you to Tom Dillon for providing feedback and support during our design and execution of our project. Additionally, we would like to thank David Sapienza for providing our team space to conduct calls with our mentors, for David Ferguson for allowing us to borrow lab equipment, Dr. Jason Dunkle for aiding us in our statistical analysis, and Bryan Winkelman for assisting us in the construction and maintenance of our website, giving us feedback, and guiding us in our blog posts. We would also like to thank Douglas County School District for the Innovation and Perkins Grant funding that provided our laboratory with research grade equipment.

REFERENCES

1. Astwood, J.D., Leach, J.N., & Fuchs, R.L. (1996, October 14). Stability of food allergens to digestion in vitro. *National Biotechnology*.1996;14:1269–1273. Retrieved 2015, October 8. [Web]
2. Balint, V. L. (2012, November 19). Egg allergy? New research finds some kids can tolerate eggs in baked goods.*Raising Arizona Kids*. Retrieved 2015, October 5. [Web]
3. Biotechnology Explorer ELISA Immuno Explorer Kit Instruction Manual. (2016). *Bio-Rad Laboratories, Inc.* Retrieved, 2015, January 21. [Print and Web]
4. Caubet J, Bencharitiwong R, Moshier E, & Godbold J. (2012, January 24). Significance of ovomucoid- and ovalbumin-specific IgE/IgG(4) ratios in egg allergy. *National Center for Biotechnology Information*. Retrieved 2015, October 21. [Web]
5. Food Allergy Research & Education. Facts and Statistics. (2014). *Food Allergy Research & Education, Inc.* Retrieved 2016, March 31. [Web]
6. Gagne, C. (2010, February). Allergy Breakthrough on Baked Milk and Egg. *Allergic Living*. Retrieved 2016, March 17. [Web].
7. How do allergies work? (2012, March 21). *OMRF*. Retrieved 2015, October 21. [Web]
8. Verhoeckx, Kitty C.M., and Vissers, Yvonne M. "Food Processing and Allergenicity." *Food and Chemical Toxicology* 80 (2015): 1-350.

ABOUT THE AUTHORS

Pictured: Our mentor, Patricia Merkel, with National Jewish Health, Analissa, Keegan, Mark, and our instructor, Shawndra Fordham (left to right). Not pictured: Dr. Vijaya Knight.

The research opportunity we have received this year has been invaluable, both the in amount of experience we have currently gained and the amazing connections we have made for the future. This class has provided us with the best possible exposure to the field of biotechnology, giving a first hand perspective on professional level lab research. We have truly learned the importance of working as a team, setting deadlines and goals, and persevering past setbacks or challenges - skills necessary both inside and outside a laboratory setting as we continue to feed our interest in science. We hope to take these new skills with us as we begin our college careers, taking advantage of every opportunity in the future.

5810 McArthur Ranch Road
Highlands Ranch, CO 80124
303-387-3000

Principal
Andy Abner
Andrew.Abner@dcsdk12.org

Registrar
Polly Poindexter
Polly.Poindexter@dcsdk12.org

Administrative Assistant
Barb Cocetti
Barbara.Cocetti@dcsdk12.org

STEM PROGRAMMING

The Principals of Experimental Design in Biotechnology course is one of many courses offered as part of our choice-driven STEM programming, which allows each of our students to prepare for their vision of a career in science, technology, engineering, or math.

Due to the competitive nature of STEM majors in college, we believe that taking a rigorous course load, including Honors, AP, and dual credit courses, is the best way to prepare students for the coursework they will encounter. In addition, involvement in clubs that encourage competition in Science, Technology, Engineering, and Business allows students the opportunities to think on their feet, construct and communicate arguments, and work through the engineering process. Finally, our wish is that students will become involved in an internship or shadowing experiences in order to gain the workplace experience that our classes may not provide.

Students who diligently pursue this tough course load, as well as meeting these additional requirements, will not only benefit from their knowledge and preparation, but will also be able to show the universities that they are determined students by presenting them with a STEM certificate.

Rock Canyon High School
Home of the Jaguars

Our Mission:
To Empower, To Explore, To Encourage and To Excel in Education

Our Vision:
Our student-centered culture practices collaborative decision making and continuous improvement in a safe, supportive environment.

Rock Canyon is a comprehensive high school consisting of grades nine through twelve, located in Highlands Ranch, Colorado, a southern suburb of Denver. Our community is composed primarily of working professionals.

Rock Canyon is part of the Douglas County School District (DCSD), the third largest school district in Colorado, serving over 67,000 students for the 2015-2016 school year. The district is comprised of 9 high schools, 9 middle schools, 47 elementary schools, 12 charter schools, 2 magnet schools, 3 alternative schools and an online school. The DCSD continues to maintain its standing as one of the finest, highest achieving districts in Colorado.

Rock Canyon opened in 2003. It has a current enrollment of 2,050 students. RCHS occupies a 279,250 square foot building on an 80-acre campus.

Rock Canyon High School prides itself on excelling in academics, activities and athletics to create a balanced and comprehensive high school experience for all students. We strive to develop a tradition of excellence in order to develop a premier high school program for all post-secondary options. Rock Canyon is currently ranked as one of the top high schools in Colorado.

We invite our parents to take an active role in their student's education by empowering their students to explore the many opportunities offered at Rock Canyon while continuing to encourage their students to excel in their educational goals. We truly believe a partnership must exist between the school and the family; together we can elevate our students to the next level.